Clay Science - Bridging Geology and Physical Chemistry - An Interdisciplinary Approach to Material Science and Earth Processes

Edited by Walid Oueslati

Published in London, United Kingdom

Clay Science - Bridging Geology and Physical Chemistry - An Interdisciplinary Approach to Material Science and Earth Processes
http://dx.doi.org/10.5772/intechopen.1008046
Edited by Walid Oueslati

Contributors
Agi Augustine Aja, Ahmed A. Khalifa, Barima Money, David Abutu, Eman M. Abdel Hamid, Hakan Sahin, Huicong Chen, Ling Liang, Mariame Coulibaly, Mirna Daye, Norasyikin Ismail, Qiwen Wu, Rayan Hassan Modather, Siti Qurratu' Aini Binti Mahat, Wenjing Wang, Zhi Hong

First published in London, United Kingdom, 2025 by IntechOpen
IntechOpen is the global imprint of INTECHOPEN LIMITED, registered in England and Wales, registration number: 11086078, 167-169 Great Portland Street, London, W1W 5PF, United Kingdom

For EU product safety concerns: IN TECH d.o.o., Prolaz Marije Krucifikse Kozulić 3, 51000 Rijeka, Croatia, info@intechopen.com or visit our website at intechopen.com.

British Library Cataloguing-in-Publication Data
A catalogue record for this book is available from the British Library

Clay Science - Bridging Geology and Physical Chemistry - An Interdisciplinary Approach to Material Science and Earth Processes
Edited by Walid Oueslati
p. cm.
Print ISBN 978-1-83635-133-7
Online ISBN 978-1-83635-132-0
eBook (PDF) ISBN 978-1-83635-134-4

If disposing of this product, please recycle the paper responsibly.

Meet the editor

Dr. Walid Oueslati is a Professor in the Physics Department at the Faculty of Sciences of Bizerte, University of Carthage, Tunisia. His research expertise spans experimental physics, materials science, and mineralogy, with particular emphasis on crystallography and X-ray powder diffraction analysis. Dr. Oueslati has established an international reputation for his innovative work on nanomaterials, clay nanofillers, and clay/polymer nanocomposites, advancing sustainable water treatment technologies. His multidisciplinary approach uniquely bridges geological research and material science, encompassing hydrogeology, groundwater science, and sustainable water management. This interdisciplinary perspective positions Dr. Oueslati as a leading researcher in both fundamental clay mineralogy and crystallography, as well as their practical applications in environmental sustainability and groundwater system investigations.

Contents

Preface

Clay minerals, which lie at the intriguing nexus of physical chemistry and geology, are among nature's most amazing materials. These incredibly adaptable materials, which were created over geological timescales through intricate weathering and transformation processes, are now crucial to solving some of humanity's most significant problems, from chemical catalysis and waste treatment to environmental monitoring and sustainable building.

The interdisciplinary nature that defines modern clay science is reflected in the cutting-edge research presented in this volume. This book's five chapters show how fundamental knowledge of clay chemistry and mineralogy is being transformed into innovative uses across numerous fields. This book emphasizes the exceptional contributions that clay materials make to current materials research through the integration of geological description and physical-chemical applications.

The book is divided into three subject sections that illustrate the logical process from technological innovation to basic characterization:

The fundamental principles of clay research are outlined in Section 1, *Clay Characterization and Properties*. The first chapter offers a thorough review of X-ray diffraction methods for identifying clay minerals, including important minerals such as smectite, kaolinite, and illite, for a variety of applications, including geotechnical engineering and Martian mineralogy. Egyptian bentonite can be transformed into a high-performance, sustainable binder through physical purification techniques, underscoring the practical significance of understanding clay characteristics for material optimization, which is described in the second chapter.

Section 2, *Environmental and Analytical Applications,* shows how the unique physical-chemical characteristics of clays can be used for chemical conversion and environmental protection. The modification of clays for electrochemical sensing demonstrates that their surface chemistry and layered structure provide an ideal basis for detecting contaminants. The versatility of natural clay materials in promoting chemical reactions is highlighted in Hong et al.'s assessment of kaolin as a catalyst in organic reactions, whether in its native or modified form.

In Section 3, *Structural Applications and Geopolymers,* clay science is expanded to include sustainable industrial innovation. In their comprehensive bibliometric analysis and literature review of oil-well cementing geopolymers made from clays, Money et al. explore how the alkaline activation of clay minerals yields materials with superior mechanical properties and environmental sustainability profiles. This work serves as an example of how future construction materials can be developed by comprehending the chemistry at the core of clay minerals.

These contributions collectively highlight a key idea: clay science is inherently multidisciplinary, requiring a seamless integration of physical-chemical concepts with geological information. The foundation for physical-chemical applications is provided by the geological origins of clays, including their formation processes, structural traits, and mineralogical diversity. On the other hand, the techniques and viewpoints of physical chemistry are necessary to comprehend the surface chemistry, ion-exchange capacity, catalytic activity, and electrochemical behavior of clays. Researchers, engineers, and students working in the field of clay science—from basic mineralogy to applied materials engineering—will find this volume useful. In addition to presenting novel research or thorough reviews, each chapter demonstrates how combining geological and physical-chemical viewpoints yields creative solutions to practical problems.

As we face worldwide imperatives toward sustainable resource use, ecological monitoring, and resource efficiency, clay minerals—ubiquitous, versatile, and eco-friendly—have great potential. The research detailed in this volume augments our fundamental understanding and provides opportunities for applications that will set the stage for future technologies.

We expect this publication to inspire continued research at the interface between geology and physical chemistry, where clay science continues to yield new avenues for sustainable innovation.

Walid Oueslati
Faculty of Sciences of Bizerte,
Physics Department,
University of Carthage,
Bizerte, Tunisia

Dedication

-In loving memory of my father, Eng. Rachid Oueslati (05/12/2024), whose wisdom, dedication, and unwavering support shaped the person I am today.

-To my beloved wife Yosra, whose patience, love, and encouragement made this work possible.

-To my children Emna, Fatma Ezzahra, and Mohamed Ali, who inspire me every day to build a better future.

Section 1

Clay Characterization and Properties

Chapter 1

Clay Mineral Identification Using X-Ray Diffraction

Mirna Daye and Hakan Sahin

Abstract

This chapter presents a comprehensive overview of clay mineral characterization using X-ray diffraction (XRD), a fundamental analytical tool widely applied in geosciences, environmental studies, and planetary science. It begins by outlining the theoretical principles of XRD, including Bragg's Law, and details essential methods for clay sample preparation, such as oriented mounts and treatments like glycolation and heating, which aid in distinguishing between major clay groups. The identification of minerals such as kaolinite, illite, smectite, and chlorite, as well as the detection of mixed layer clays, is thoroughly discussed. The chapter also explores the role of XRD in monitoring mineralogical changes during diagenesis, and microbial interactions. In addition, it highlights the importance of clay mineral detection in Martian surface exploration, where XRD analyses, such as those conducted by the Curiosity rover offer insights into past aqueous activity and habitability. By integrating terrestrial and extraterrestrial case studies, this chapter underscores the broad applicability of XRD in clay science. Building on this foundation, it further examines how XRD-derived mineralogical data inform geotechnical engineering practices, including expansive soil characterization, soil–water behavior modeling, stabilization design, and long-term performance monitoring. Applications in offshore geotechnics, environmental containment, foundation interaction modeling, and GIS-integrated remote sensing are also addressed. These engineering perspectives demonstrate how mineral-level precision enhances design reliability, risk assessment, and infrastructure resilience in clay-rich environments. XRD enables early detection of problematic mineral phases, supports performance-based treatment selection, and improves the predictive accuracy of mechanistic models used in infrastructure design.

Keywords: clay minerals, X-ray diffraction (XRD), smectite, kaolinite, illite, expansive soils, soil stabilization, geotechnical engineering, soil–water characteristic curve (SWCC), mineralogical analysis, environmental containment, offshore geotechnics, remote sensing, infrastructure resilience, Martian mineralogy

1. Introduction

Clay minerals are fundamental components of soil and sedimentary environments, influencing a wide range of geotechnical, environmental, and geological processes. Their unique layer structures, variable chemical compositions, and surface properties govern critical behaviors such as swelling, shrinkage, permeability, shear strength, and contaminant retention. As a result, accurate identification and quantification of clay minerals are essential for predicting soil performance and designing effective engineering solutions.

Among the various mineral identification techniques, X-ray diffraction (XRD) stands out as the most definitive and widely applied method for characterizing crystalline phases in clays. XRD provides direct insight into mineral structures, enabling the detection of key diagnostic features such as basal spacings, peak intensities, and thermal stability. These characteristics are essential for distinguishing between mineral groups like smectite, kaolinite, illite, and mixed-layer clays. This analytical capability extends across disciplines, supporting both scientific investigations into clay genesis and applied engineering analyses for infrastructure design, hazard mitigation, and environmental management.

This chapter is organized into two complementary sections. Section 2 presents the scientific basis and analytical methods for clay mineral identification using XRD, covering diffraction principles, sample preparation, and pattern interpretation within a geological and geochemical context. Section 3 builds upon this foundation to explore engineering perspectives, illustrating how XRD-derived mineralogical data are integrated into geotechnical design, soil stabilization, performance modeling, and long-term monitoring. Together, these sections provide a comprehensive view of how mineralogical characterization using XRD bridges fundamental science and practical engineering.

2. Scientific basis and analytical methods for clay mineral identification using X-ray diffraction (XRD)

X-ray diffraction (XRD) is a fundamental technique for determining the mineralogical composition of soils, sediments, and rock materials in geological and geochemical studies. By analyzing the diffraction patterns generated when X-rays interact with a crystalline lattice, it is possible to determine structural parameters such as interlayer spacings, symmetry, and degree of crystallinity. These parameters allow clear differentiation between smectite, kaolinite, illite, and mixed-layer clays.

This section outlines the principles of diffraction and Bragg's Law, along with key considerations for sample preparation and treatment. Emphasis is placed on interpreting diagnostic peak positions, intensities, and shifts that provide accurate clay mineral identification. These analytical foundations support the engineering applications presented in later sections.

2.1 Principles of X-ray diffraction for clay minerals

X-ray diffraction (XRD) is based on the constructive interference of monochromatic X-rays with a crystalline sample. When X-rays interact with a crystalline lattice, they are diffracted at specific angles in accordance with Bragg's Law: $n\lambda = 2d \sin\theta$. Each clay mineral has a characteristic set of basal spacing values that enable its identification [1, 2]. For instance, kaolinite is characterized by peaks at approximately 7.1 Å for the (001) plane and 3.57 Å for the (002) plane. Illite typically exhibits a strong reflection near 10 Å. Smectite displays an expansion from around 12–15 Å to approximately 17 Å upon ethylene glycolation. Chlorite, on the other hand, is recognized by its distinctive peaks near 14 Å and 7 Å. These diffraction patterns provide a reliable basis for the mineral specific identification of clay phases.

2.2 Sample preparation

Proper sample preparation is critical for obtaining reliable XRD results when analyzing clays. Typically, clays are separated from the bulk rock or soil by isolating the <2 μm fraction [2]. Disaggregation is carried out by dispersing the sample in distilled water with sodium hexametaphosphate (commonly ~5 g/L) as a deflocculant. Particle-size separation is then achieved by settling based on Stokes' Law to isolate the clay-size fraction. The separated clay suspension is used to prepare oriented mounts on glass slides, which enhance basal reflections and improve mineral identification. Depending on the experimental design, slides are either air-dried at room temperature or oven-dried at ~40–50°C to maintain clay structural integrity. To further distinguish between clay minerals such as smectite, illite, chlorite, and kaolinite, additional treatments are applied: air-drying, ethylene glycolation, and heating to 550°C [3].

2.3 Interpretation of XRD patterns

Interpreting XRD patterns involves recognizing the diagnostic basal spacings of clay minerals and understanding how these spacings change under various treatments. Kaolinite typically exhibits a sharp basal reflection at ~7.15 Å that remains unchanged after ethylene glycolation or heating to 550°C. Illite displays a prominent peak near 10 Å, also stable under ethylene glycolation and heating, though partial collapse can occur at higher temperatures if interlayer cations are altered. Smectites, such as montmorillonite, show a basal reflection near 12–15 Å in air-dried conditions, which expands to ~17 Å upon ethylene glycolation and collapses to ~10 Å after heating to 550°C. Chlorite is identified by its basal peak near 14.2 Å, which remains stable under ethylene glycolation but exhibits a distinct 7.1 Å peak that persists after heating, distinguishing it from kaolinite.

As illustrated in **Figure 1** (*adapted from* [2]), the diagnostic basal spacings for common clay minerals can be identified by their characteristic diffraction peaks under different treatments. Careful comparison of observed peaks with reference patterns, in combination with sample preparation history, ensures accurate mineral identification and helps prevent misinterpretation caused by mixed-layer minerals or preferred orientation effects.

2.4 Applications in geobiology and geochemistry

Clay minerals play a central role in geobiological and geochemical processes due to their high surface area, reactivity, and ability to sorb cations, organic matter, and trace metals. X-ray diffraction (XRD) is essential for identifying mineralogical changes that reflect environmental conditions, microbial activity, and diagenetic transformations.

2.4.1 Paleoenvironments and weathering indicators

Clay mineral assemblages serve as sensitive indicators of paleoclimatic conditions and weathering regimes [4]. For instance, a high abundance of kaolinite typically reflects intense chemical weathering in warm, humid climates [5]. In contrast, illite is commonly associated with physical weathering processes and the presence of cooler climatic conditions. Smectite, on the other hand, often forms in

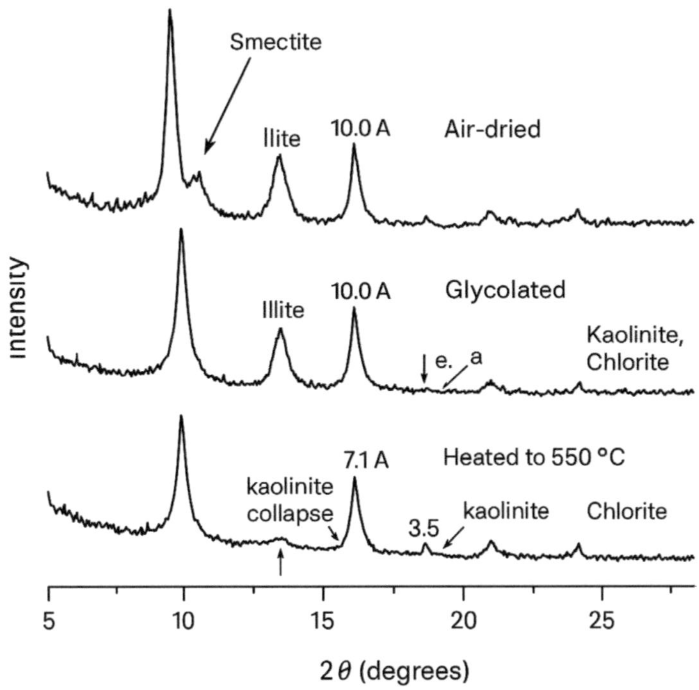

Figure 1.
X-ray diffraction (XRD) patterns of clay samples under different treatments: air-dried, glycolated with ethylene glycol (EG), and heated to 550°C. Diagnostic basal reflections are shown for common clay minerals: the ~12 Å peak that expands to ~17 Å after EG treatment confirms the presence of smectite; the stable 10 Å reflection indicates illite; the 7.1 Å peak that collapses upon heating identifies kaolinite; and the persistence of a 7 Å peak after heating indicates chlorite. The label "a" refers to the basal spacing (d-spacing) of the strongest (001) reflection in angstroms (Å), while "e" denotes ethylene glycol treatment. These phase changes reflect interlayer expansion and thermal stability, which are key criteria for clay mineral identification.

semi-arid environments characterized by poor drainage. XRD analysis of sediment cores enables reconstruction of past environmental changes, especially in lacustrine and marine contexts, providing insight into Quaternary climate fluctuations and provenance.

2.4.2 Microbe: Clay interactions

In geobiology, interactions between microorganisms and clay minerals play a key role in mineral transformations, redox cycling, and biosignature formation in both modern and ancient environments. Iron-rich clays, like nontronite (Fe(III)-smectite) are highly susceptible to microbial reduction under anoxic conditions.

Dissimilatory iron-reducing bacteria (DIRB), such as *Shewanella oneidensis* and *Geobacter sulfurreducens*, oxidize organic substrates while reducing structural Fe(III) in clays. Thereby using the mineral itself as an electron acceptor [6–9]. This process produces Fe(II), which can alter basal spacings (e.g., ~14 Å to ~10 Å), lower crystallinity, and over time drive transformations into secondary minerals such as illite or chlorite [10–12].

These microbial activities leave morphological, structural, and chemical imprints within clays that may persist over geologic timescales. For example, nanoscale lattice changes and mixed-layered phases can be detected using advanced tools like synchrotron-based XRD and TEM [6, 13]. Such transformations also modify clay surface reactivity, influencing their capacity to adsorb and retain organic matter and thereby enhancing biosignature preservation [14, 15].

Beyond geologic records, these interactions have applied significance. In environmental remediation, DIRB-mediated reduction can immobilize heavy metals and radionuclides [16]. In planetary science, understanding microbe–clay systems inform the interpretation of Martian phyllosilicates and their potential to record traces of ancient life.

2.4.3 Diagenesis and hydrocarbon reservoirs

Clay minerals are highly sensitive to diagenetic transformations that occur during burial. One of the most significant of these is the gradual transformation of smectite into illite, known as illitization. This reaction is strongly influenced by factors such as temperature, potassium availability, and the duration of burial. The smectite to illite transition plays a critical role in petroleum systems, as it directly impacts reservoir properties including porosity, permeability, and the development of overpressure zones [17]. Mixed-layer clays, particularly illite–smectite (I–S) interstratifications, can be analyzed and quantified through XRD techniques to evaluate the thermal maturity of sedimentary basins [3].

2.4.4 Element cycling and environmental geochemistry

Clays play a pivotal role in elemental cycling within soils and sediments due to their capacity to adsorb and exchange a variety of nutrients and contaminants. Their surfaces are highly reactive, allowing for the retention and exchange of essential nutrients such as potassium (K^+) and ammonium (NH_4^+), as well as toxic pollutants like lead (Pb^{2+}) and arsenic (As^{3+}). X-ray diffraction (XRD) serves as a powerful analytical tool to monitor the mineralogical changes that occur during environmental processes, including acid mine drainage, waste stabilization, and interactions between clays and organic matter in soils. For example, smectite, known for its high cation exchange capacity, is particularly effective at binding heavy metals and organic contaminants. Its structural stability and capacity for contaminant retention following environmental treatments can be effectively evaluated through XRD analysis.

2.4.5 Microbial biosignatures in ancient clays

Clay minerals, especially Fe-Mg smectites like nontronite, are especially effective at trapping and preserving organic matter. Their very fine grain size, large surface area and swelling interlayers allow them to adsorb and intercalate organic molecules and metal ions, physically protecting organics from degradation. In ancient sediments and paleosols, layers rich in Fe/Mg–smectite clays often coincide with higher preserved organic carbon [18, 19]. The redox state of a paleosol is the main control on organic retention, with chemically reduced horizons enriched in iron–smectites and amorphous material being especially favorable for biosignatures [18]. In short, Fe-rich clays combine high reactivity with chargeable surfaces, making them excellent matrices for incorporating and preserving microbial organic residues [18, 19].

Furthermore, microbes can transform clay minerals in ways that leave lasting traces. Dissimilatory Fe(III)-reducing bacteria (e.g. *Shewanella*, *Geobacter*) readily use structural Fe(III) in clays as an electron acceptor, producing Fe(II) and breaking down the clay lattice [16]. Laboratory experiments show that *Shewanella* can reduce a large fraction of the structural Fe(III) in smectite or illite, causing layer collapse and alkali/Al enrichment [20]. This "microbial reductive dissolution" of clays often generates new phases: in situ bioreduction of illite, for example, releases Fe^{2+} into porewaters even at low temperature. Jung *et al.* [21] demonstrated that microbes beneath the Larsen Ice Shelf break down clay-bound Fe(III) and liberate dissolved Fe^{2+}, effectively altering the clay's structure and chemistry [21]. Over time, such reductive processes can drive recrystallization or "illitization" of smectites, as Fe is removed and K/Al in the layers becomes relatively enriched [20].

The mineralogical and geochemical fingerprints of these processes can be preserved as biosignatures. In anoxic, iron-rich sediments, microbial Fe reduction (often coupled to fermentation or methanogenesis) leads to authigenic Fe^{2+}-minerals such as magnetite, siderite ($FeCO_3$) and vivianite ($Fe_3(PO_4)_2{\cdot}8H_2O$). Recent work in a ferruginous lake analog (Lake Towuti) shows that framboidal magnetites (primary precipitates) become overgrown by acicular Ni-sulfides (millerite) and by siderite and vivianite during burial, all driven by microbial redox gradients. These phases act as biosignatures of microbial iron and sulfate reduction, fermentation and methanogenesis. In ancient clay-rich rocks (e.g. Archean paleosols or BIFs), such authigenic minerals, together with iron isotopic shifts or enrichments of redox-sensitive trace metals (V, Mo, U) at redox fronts, would mark early microbial activity. In summary, iron-bearing clays not only incorporate organic biosignatures, but their microbially induced alteration products (collapsed lattices, Fe^{2+}-phases, isotopic anomalies) provide a suite of mineralogical and geochemical signatures of past life [18, 19, 22].

2.4.5.1 Mechanisms of biosignature formation

Dissimilatory iron-reducing bacteria (DIRB), particularly species of *Shewanella* and *Geobacter*, play a key role in transforming Fe-bearing clay minerals through redox reactions that leave behind distinct mineralogical and geochemical signatures. These microbes are capable of utilizing structural Fe(III) in the octahedral sheets of smectites such as nontronite as terminal electron acceptors during anaerobic respiration [6, 8, 23]. Upon microbial reduction, Fe(III) is converted to Fe(II), which alters the mineral structure and destabilizes the original lattice. This process frequently results in a measurable collapse of the basal spacing (as detected by X-ray diffraction), along with peak broadening or shifting, indicating clay lattice contraction and mineral reconfiguration [6, 12].

These structural changes are often preserved as biosignatures. Lattice distortions, especially contractions in d-spacing, serve as mineralogical fingerprints of past Fe reducing microbial activity [6, 23]. In some cases, repeated microbial–geochemical cycling can result in interstratification or mixed-layering of clays [3], producing diagnostic XRD patterns not attributable to purely abiotic alteration. Moreover, trace element enrichments (e.g., Mo, Ni, V) that colocalize with Fe reduction zones are increasingly recognized as geochemical biosignatures of microbial metabolism [15]. These trace metals may become incorporated into secondary minerals or adsorbed onto clay surfaces in microbially active zones.

At the microscale, microbial alteration of clay minerals can produce heterogeneity such as zoning, overgrowths, and microtextures surrounding former cell locations. Bishop et al. [24], using high resolution imaging and spectroscopy, observed mineral coatings and zoned clays enveloping fossilized microbial filaments, supporting the idea that biological processes directly influence mineral morphology during diagenesis [24]. Such microtextures can be exceptionally well preserved in fine-grained clays, especially under rapid burial or anoxic conditions. Altogether, microbial Fe reduction leaves behind a suite of structural, chemical, and textural markers that can function as preserved biosignatures in both modern analogs and ancient clay-rich deposits.

2.4.5.2 Techniques for detection

X-ray diffraction (XRD) is a primary method for detecting microbial alteration in clays by identifying reductions in basal spacings, such as the shift from 14 Å to 10 Å in smectite upon iron reduction, quantifying changes in crystallinity index and mixed layered phases, and monitoring mineral transitions like smectite to illite, which are often triggered or accelerated by microbial processes in diagenetic environments [3]. Other complementary techniques include transmission electron microscopy (TEM), which visualizes nanoscale microbial imprints and mineral textures [6]; Mössbauer spectroscopy, which determines Fe(II)/Fe(III) ratios [23]; Raman and FTIR spectroscopy, which detect bio-organic compounds associated with clay minerals [15]; and stable isotope analysis, which measures $\delta^{13}C$ and $\delta^{34}S$ anomalies linked to microbial metabolism [25].

2.4.5.3 Significance for early earth and astrobiology

Precambrian clay-rich formations, such as the Doushantuo (China) and Gunflint (Canada), preserve microbial biosignatures including mineralized microfossils, isotopic anomalies, and clay fabric alterations [6, 15]. On Mars, smectite-bearing regions like Mawrth Vallis and Nili Fossae are key astrobiological sites: they formed under aqueous, potentially habitable conditions, can sorb and protect organic matter, and may retain mineralogical evidence of ancient microbial activity [24, 25].

2.4.5.4 Martian rover missions and clay biosignatures

Mars has long been a focus in the search for extraterrestrial life due to geological evidence of past water activity. Clay minerals (phyllosilicates), in particular, are critical because they form in aqueous environments and have high potential to preserve organic molecules and microbial biosignatures. Recent rover missions, especially NASA's Curiosity (Mars Science Laboratory) and ESA's upcoming Rosalind Franklin rover, have targeted clay-rich terrains for this reason.

The Curiosity rover, since its landing in 2012, has explored Gale Crater, a 154 km-wide impact basin that once hosted a lake. X-ray diffraction (XRD) data from the CheMin instrument aboard Curiosity confirmed the presence of smectite clays in the Murray formation and Vera Rubin Ridge, suggesting long-lasting lacustrine conditions [26, 27].

The detection of Fe/Mg-rich smectites, Amorphous silica (a biosignature preserving phase), and organic molecules, such as thiophenes and chlorobenzene [28], suggests that Gale Crater once harbored habitable environments with clay minerals potentially preserving signs of past microbial life.

The Rosalind Franklin rover, part of ESA's ExoMars mission, is scheduled to land at Oxia Planum, a site characterized by extensive exposures of Fe/Mg phyllosilicates, likely formed in neutral to mildly alkaline waters [29]. This landing site hosts layered clay rich deposits, spectral signatures of nontronite, an Fe rich smectite and geomorphological evidence of ancient surface hydrology, including deltaic formations and inverted channels. Equipped with advanced instruments such as the Mars Organic Molecule Analyzer (MOMA), Raman Laser Spectrometer (RLS), and MicrOmega infrared hyperspectral microscope, the rover will analyze subsurface samples obtained by a 2-meter drill, the deepest ever used on Mars [30]. These instruments will differentiate clay phases in situ, detect preserved biomolecules shielded from surface radiation, and map associations between minerals and organics, which are crucial for authenticating biosignatures [31]. The biosignature preservation potential in Martian clays is supported by their capacity to adsorb and protect organic molecules from degradation [32], their formation in environments potentially analogous to microbial habitats on early Earth [14], and their presence in geologic units dating to the Noachian–Hesperian transition, a period hypothesized to be favorable for life. Together with high-resolution orbital mineralogical data, such as those from CRISM, the ExoMars mission represents a significant step toward identifying ancient biosignatures on Mars.

2.4.5.5 Methodologies for detecting biosignatures in returned Martian samples

The prospect of returning Martian samples to Earth offers a unique opportunity to apply the full suite of advanced analytical techniques to search for biosignatures: molecular, isotopic, or mineralogical evidence of life or biological processes. These analyses can target clay-rich samples, which are prime candidates for biosignature preservation due to their high surface area, cation exchange capacity, and ability to sorb organic matter.

2.4.5.5.1 Mineralogical and structural analysis

Synchrotron-based X-ray diffraction (SXRD) offers ultra-high resolution phase identification and crystallographic analysis, enabling the detection of subtle microbial alterations in clay lattices, such as lattice spacing reductions caused by Fe-reduction. Beyond phase identification, SXRD can reveal micro strain variations and shifts in peak positions that reflect cation exchange or removal during microbial activity, providing insights into reaction mechanisms at the atomic scale. Transmission Electron Microscopy (TEM) provides nanometer-scale imaging of mineral–organic associations, including cell shaped morphologies, lattice defects, and nanocrystalline domains that may represent potential microbial imprints [6]. TEM can also resolve the interface between microbial EPS and mineral surfaces, revealing nucleation sites and alteration fronts. Cryo-TEM and Cryo-SEM techniques preserve the hydrated state of clays and biofilms during imaging, thus avoiding dehydration artifacts and maintaining the integrity of delicate microstructures. These cryogenic methods also allow for the visualization of transient mineral phases and labile organic coatings that would otherwise be lost during conventional preparation, offering a more accurate representation of in situ microbe–mineral interactions.

2.4.5.5.2 Organic molecule detection

Gas Chromatography–Mass Spectrometry (GC–MS) is a powerful technique for detecting and identifying organic compounds such as lipids, thiophenes, and amino acids that may be of biological origin. By separating complex mixtures and providing mass spectral fingerprints, GC–MS enables both qualitative and quantitative assessments of biomarker distributions. Derivatization agents like TMAH (Tetramethylammonium hydroxide) and MTBSTFA (N-Methyl-N-(tert-butyldimethylsilyl)trifluoroacetamide) are commonly used to extract organics bound to mineral surfaces or trapped within mineral matrices, enhancing detection sensitivity and enabling the analysis of otherwise recalcitrant compounds. These derivatization steps also improve compound volatility and stability, facilitating their analysis via GC–MS without significant thermal degradation. Fourier Transform Ion Cyclotron Resonance Mass Spectrometry (FT-ICR-MS) offers ultra-high-resolution analysis of complex organic mixtures, resolving thousands of molecular formulas within a single sample. This capability allows for the differentiation of abiotic versus biotic sources based on molecular complexity, heteroatom content, and compound class distributions [28]. When combined, GC–MS and FT-ICR-MS provide a complementary toolkit for reconstructing the diversity, origin, and preservation state of organic molecules associated with microbe–mineral systems, offering critical insights into past and present biosignatures.

2.4.5.5.3 Isotopic analysis

Stable Isotope Ratio Mass Spectrometry (IRMS) measures isotopic compositions such as $\delta^{13}C$, $\delta^{15}N$, and $\delta^{34}S$ with high precision, often capturing characteristic biological fractionation patterns. Light $\delta^{13}C$ signatures in carbonates or organic matter, for instance, can indicate carbon fixation pathways driven by microbial activity. Similarly, $\delta^{15}N$ and $\delta^{34}S$ variations can reveal nitrogen and sulfur cycling processes, respectively, that are frequently mediated by microbial metabolisms. Secondary Ion Mass Spectrometry (NanoSIMS) extends these capabilities by providing micron- to submicron-scale spatial resolution of isotopic distributions across mineral–organic interfaces. This enables the direct association of isotopic anomalies with specific microstructures, such as microbial cell outlines or EPS-rich regions, effectively linking geochemical signals to morphological biosignatures. When used together, IRMS and NanoSIMS provide a powerful combination for both bulk and spatially resolved isotopic characterization, allowing researchers to identify, localize, and interpret biosignatures with a high degree of confidence.

2.4.5.5.4 Spectroscopy techniques

Raman Spectroscopy identifies carbonaceous matter, mineral phases, and potential biological pigments or structural biomolecules. Its ability to detect the degree of structural order in carbon, particularly through the analysis of the D/G band ratio, can provide strong evidence for biogenic carbon when combined with morphological context. This non-destructive technique is especially valuable for studying delicate microfossils or mineral-organic associations in situ. X-ray Absorption Near-Edge Spectroscopy (XANES) determines the oxidation states and coordination environments of elements such as iron, sulfur, and carbon, offering detailed insights into past microbial metabolisms and redox-sensitive changes associated with biological activity. When applied to microbe–mineral systems, XANES can reveal subtle chemical fingerprints of biotic interactions, complementing morphological and isotopic data to strengthen biosignature interpretations.

2.4.5.5.5 Microscopy and imaging

Focused Ion Beam (FIB)–Scanning Electron Microscopy (SEM) enables the precise slicing of clay aggregates and microfossils at the nanometer scale, facilitating the preparation of site-specific thin sections for detailed isotopic or spectroscopic analysis. This technique allows researchers to target specific regions of interest identified through preliminary imaging, ensuring that subsequent analyses focus on features with the highest potential for biosignature preservation. Three-dimensional tomography techniques, such as ptychographic X-ray computed tomography, enable the non-destructive reconstruction of organic mineral structures without physical sectioning. This preserves delicate and fragile features that might otherwise be lost or altered during sample preparation, providing unprecedented insight into the spatial arrangement and connectivity of biosignature bearing phases.

2.4.5.5.6 Biomolecule detection

Proteomics and genomics (including metagenomics) approaches can be applied when preserved DNA, RNA, or peptides are present within the sample. Techniques such as next-generation sequencing (NGS) and peptide mass fingerprinting can help identify microbial lineages, metabolic pathways, and functional traits linked to past or present biosignatures. Given the extremely low biomass often encountered in ancient or extraterrestrial samples, contamination control is crucial at every stage, from sample acquisition to laboratory processing, to ensure authenticity of the detected biomolecules [33]. These methods, when integrated with mineralogical, isotopic, and organic geochemical analyses, can provide a comprehensive understanding of the biological and environmental context of the studied systems.

Returned Martian samples will be analyzed in biosafety level containment laboratories with Earth based sterilization controls and blanks. Instrument protocols will be validated against Mars analog materials (e.g., Atacama clays, hydrothermal sediments) to distinguish Martian signals from terrestrial contaminants.

2.4.5.5.7 Limitations and complementary methods

X-ray diffraction (XRD) has certain limitations: it cannot detect amorphous or poorly crystalline materials, and mixed-layer clays often require complex modeling and profile fitting to interpret their diffraction patterns accurately [3]. Therefore, XRD analyses should be complemented with techniques such as Fourier Transform Infrared Spectroscopy (FTIR), Scanning Electron Microscopy with Energy Dispersive Spectroscopy (SEM-EDS), or Transmission Electron Microscopy (TEM) to obtain detailed information on morphology and elemental composition [17].

Advances in spectroscopic tools like Raman and infrared techniques have augmented the study of clay minerals, enabling precise detection of both crystalline and amorphous phases. Raman spectroscopy, for instance, facilitates the identification of vibrational modes unique to specific clays, while infrared spectroscopy provides insights into hydroxyl bonding and the presence of organic residues. These methods complement XRD and can verify ambiguous findings, particularly in samples with complex mineralogical compositions. In planetary exploration, these methods are invaluable for probing extraterrestrial soils, such as Martian regolith, where mixed signals from minerals and potential organics demand nuanced interpretation.

3. Engineering perspectives in clay mineral identification using X-ray diffraction (XRD)

Engineering applications of X-ray diffraction (XRD) extend beyond mineralogical classification to the practical evaluation and design of infrastructure in environments containing clay-rich soils. The mineralogical composition identified through XRD directly influences engineering properties such as strength, compressibility, swelling potential, and permeability. These characteristics govern the performance of pavements, foundations, retaining structures, earthworks, and containment systems.

This section presents the integration of XRD results into engineering analyses, focusing on their role in expansive soil characterization, prediction of soil–water characteristic behavior, calibration of mechanistic–empirical design models, and evaluation of stabilization treatments. It also addresses the use of XRD in environmental containment, offshore geotechnics, and the long-term monitoring of mineralogical stability in engineered systems. The discussion emphasizes how mineralogical precision obtained from XRD enhances the accuracy, reliability, and durability of engineering designs.

3.1 XRD in expansive soil characterization

Expansive soils, particularly those dominated by smectite-group minerals such as montmorillonite, exhibit significant volumetric changes in response to moisture fluctuations. These changes can cause severe distress in pavements, shallow foundations, retaining structures, and slopes [34, 35]. X-ray diffraction (XRD) provides the most reliable means of identifying these swelling-prone minerals by detecting their diagnostic basal spacings, which expand from ~12–15 Å (air-dried) to ~17 Å after ethylene glycol treatment [2].

The direct quantification of expansive mineral content enables engineers to predict heave potential more accurately than index tests such as Atterberg limits, which may underestimate swelling risk when plasticity index values are moderate but smectite content is high [36]. For example, in the Dallas–Fort Worth region, XRD confirmed montmorillonite dominance in subgrades linked to differential heave in residential foundations [35]. This mineralogical insight allows for the selection of mitigation strategies, including moisture barriers, soil replacement, or lime stabilization. These strategies should be implemented only after confirming mineral compatibility to avoid deleterious reactions such as sulfate-induced heave [37]. **Table 1** summarizes the engineering implications of major clay minerals, including their typical basal spacings, cation exchange capacities (CEC), swelling potentials, and common engineering concerns, as compiled from mineralogical and geotechnical sources [2, 38, 39].

Mineral	Typical basal spacing (Å)	CEC (meq/100 g)	Swelling potential	Common engineering concerns
Smectite/ Montmorillonite	12–15 (air-dried), 17 (glycolated)	80–150	Very high	Pavement heave, foundation movement, slope instability
Illite	10	10–40	Low–moderate	Moderate shrink–swell, frost susceptibility
Kaolinite	7	3–15	Very low	Low plasticity, good stability in construction

Table 1.
Engineering implications of major clay minerals.

3.2 Integration of XRD with soil–water characteristic curves (SWCC) and suction models

The shape and position of the Soil–Water Characteristic Curve (SWCC) are strongly influenced by clay mineralogy. Smectite-rich soils typically exhibit higher air-entry values, prolonged moisture retention, and greater hysteresis between drying and wetting cycles than kaolinitic soils [40, 41]. These characteristics directly impact shear strength, volume change, and resilient modulus in unsaturated soil mechanics [39, 42].

In engineering practice, XRD-derived mineral profiles are integrated into SWCC fitting models to improve parameter estimation and reduce model error. This approach has proven essential in mechanistic–empirical pavement design, where suction–moisture–deformation relationships govern subgrade performance [40, 43, 44]. By calibrating SWCCs with mineralogy-verified suction behavior, designers can better predict seasonal deformation, edge lift, and center lift in expansive soil environments. **Figure 2**. Soil–Water Characteristic Curves (SWCC) illustrating the influence of clay mineralogy on unsaturated soil behavior. Data and conceptual model (adapted from Sahin [40]).

3.3 Soil stabilization design and risk assessment

Chemical stabilization, commonly performed using lime or cement, is a widely used method for improving the engineering properties of clay soils. However, the effectiveness and long-term durability of stabilization depend heavily on the initial mineralogical composition.

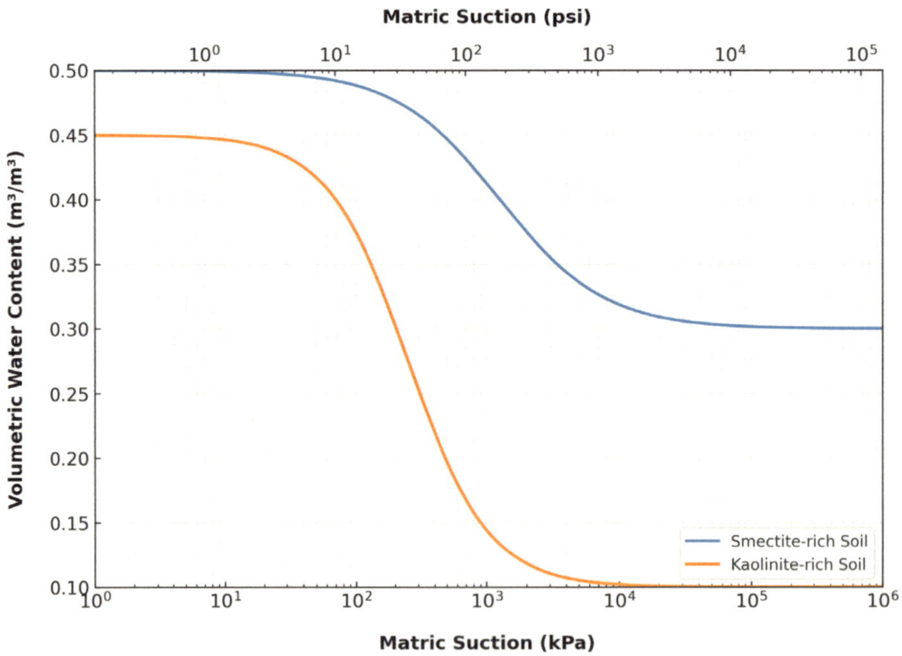

Figure 2.
Soil–Water Characteristic Curves (SWCC) illustrating the influence of clay mineralogy on unsaturated soil behavior. Data and conceptual model.

Smectite-rich soils often respond well to lime treatment, forming stable calcium silicate hydrates (CSH). In contrast, soils containing appreciable sulfates may generate expansive ettringite, which can lead to post-construction heave [37].

XRD plays a crucial role in stabilization design by identifying both reactive clay minerals and deleterious phases such as gypsum before treatment [45]. After treatment, XRD can confirm the formation of target hydration products and determine whether expansive minerals remain, enabling performance-based quality assurance. This mineralogical verification is critical for avoiding premature failures and for ensuring that the stabilization method aligns with site-specific geochemical conditions.

3.4 Cation exchange capacity (CEC) and environmental engineering applications

Clay minerals exhibit a wide range of Cation Exchange Capacity (CEC) values, with smectites and vermiculites typically possessing significantly higher CECs than kaolinites [38]. High-CEC clays are capable of retaining nutrients, heavy metals, and organic contaminants, making them highly effective in environmental containment systems such as landfill liners, mine tailings covers, and contaminant barriers.

XRD supports environmental engineering by confirming the mineral phases responsible for high CEC and by monitoring changes in mineralogy that could reduce contaminant retention over time. For example, in geosynthetic clay liners used for waste management, XRD verification of bentonite purity is essential to ensure long-term containment performance. Additionally, XRD can detect mineralogical transformations such as the degradation of smectite, which may compromise barrier integrity under chemical, thermal, or mechanical stress.

3.5 Offshore and subsea geotechnical applications

In offshore engineering, XRD is widely applied to characterize seabed sediments beneath subsea structures, mudmats, and pipelines [46]. Identifying clay minerals prone to strain softening, such as certain smectites, is essential for assessing cyclic degradation under wave and current loading.

For suction caisson and pile foundations, mineralogy determined by XRD helps predict consolidation rates, creep behavior, and cyclic strength degradation, resulting in more accurate load–displacement modeling. In subsea pipeline design, knowledge of the clay mineral composition supports the assessment of upheaval buckling risks and potential changes in embedment resistance throughout the service life. Such mineralogical insights are particularly valuable for high-consequence offshore assets where maintenance access is limited.

3.6 Foundation and structural interaction Modeling

Expansive clay mineralogy significantly influences both vertical and lateral pressures acting on foundations and retaining structures. In high-plasticity clays, lateral swelling pressures can reach 10 to 20 times the active earth pressures estimated using classical earth pressure theories [47].

Finite element models that incorporate XRD-verified mineralogical data, as demonstrated in Bulut's [48] slab-on-grade analyses, simulate moisture-induced stress changes and edge uplift more realistically than models that rely solely on index properties [48]. This mineralogy-based modeling approach enhances the reliability of designs for drilled shafts, retaining walls, and slabs-on-grade in expansive soil environments.

3.7 Laboratory protocols and best practices for engineering use of XRD

Reliable mineral identification depends on rigorous sample preparation and analysis. Best practices include isolating the <2 μm clay fraction, preparing oriented mounts, and applying sequential treatments such as air drying, ethylene glycolation, and heating to distinguish smectite, kaolinite, and illite [49].

For engineering applications, XRD results should be cross-validated with complementary techniques such as Scanning Electron Microscopy with Energy Dispersive Spectroscopy (SEM–EDS) to obtain morphological and compositional data, or Differential Thermal Analysis (DTA) to assess thermal stability. This integrated, multi-method approach ensures that mineralogical data are robust and suitable for direct incorporation into geotechnical design and modeling.

3.8 Durability and Long-term performance monitoring of clay-rich soils

The long-term engineering performance of clay-rich soils depends not only on their initial mineralogy but also on how that mineralogy evolves after construction. Environmental factors such as wet–dry cycling, freeze–thaw action, chemical leaching, and biological activity can trigger transformations such as the smectite-to-illite reaction, which can reduce swelling potential but may also lead to strength loss [3, 17].

XRD enables the detection of gradual changes in mineral assemblages during the service life of a structure. In transportation engineering, periodic XRD testing of subgrades has been used to identify lime carbonation and the loss of cementitious phases, which correlate directly with reductions in modulus and increased rutting risk [37]. Similarly, in earth dam cores, XRD monitoring has detected the dissolution of expansive phases under alkaline seepage conditions, allowing preventive maintenance before deformation becomes critical [50].

3.9 Integration of XRD with remote sensing and GIS for large-scale mapping

Hyperspectral remote sensing enables regional mapping of clay mineral distributions through spectral absorption features near 2.2 μm. However, spectral data alone cannot always distinguish similar phyllosilicates (e.g., kaolinite vs. halloysite) without ground validation. XRD provides the definitive mineralogical confirmation for remote sensing classifications, enhancing the accuracy of geotechnical hazard maps [51, 52].

In practice, this integration is invaluable for early-stage project planning. For example, pipeline route selection in Australia has used hyperspectral imagery combined with GIS, then verified with targeted XRD sampling to avoid swelling clay zones [53]. Transportation agencies have similarly combined airborne mineral maps with XRD-verified ground surveys to tailor pavement designs for moisture-susceptible subgrades.

3.10 XRD in heritage conservation and archaeological geotechnics

XRD also plays a role in the preservation of heritage structures built with earthen materials such as adobe, cob, and ancient mortars. These materials depend on specific clay mineral assemblages for cohesion and durability, which can be altered over centuries due to weathering or incompatible restoration methods [54].

In archaeological geotechnics, XRD is used to match repair materials to the original mineralogical profile, ensuring chemical compatibility and avoiding deleterious reactions. For instance, in restoring the Great Mosque of Djenné, Mali, XRD confirmed that local kaolinite-rich clays were the primary binder, guiding the selection of restoration soils to maintain structural stability under seasonal wetting–drying cycles [55]. This application demonstrates the cross-disciplinary importance of engineering mineralogy in both modern infrastructure and cultural heritage preservation.

4. Summary

X-ray diffraction (XRD) provides a direct and reliable means of identifying clay minerals, offering valuable insights that extend beyond laboratory analysis into practical engineering applications. Its precise mineralogical characterization supports geological and geochemical investigations while directly informing geotechnical design, performance prediction, and long-term maintenance strategies.

By integrating XRD data into engineering models, practitioners can better anticipate soil behavior, optimize stabilization measures, and improve the durability of infrastructure founded in clay-rich environments. The continued development of XRD techniques, coupled with complementary analytical methods, will further enhance its role in addressing complex geotechnical and environmental challenges. As a result, XRD remains a critical bridge between fundamental mineral science and engineering practice, contributing to more resilient and sustainable infrastructure systems.

Author details

Mirna Daye[1*] and Hakan Sahin[2]

1 Laboratoire LASIR UMR – CNRS, Université Lille 1 Equipe Chimie Analytique et Marine, Villeneuve d'Ascq Cedex, France

2 Department of Construction Science, Texas A&M University, College Station, Texas, USA

*Address all correspondence to: mirnadaye@gmail.com

References

[1] Biscaye PE. Mineralogy and sedimentation of recent deep-sea clay in the Atlantic Ocean and adjacent seas and oceans. Geological Society of America Bulletin. 1965;**76**(7):803-832

[2] Moore DM, Reynolds RC. X-Ray Diffraction and the Identification and Analysis of Clay Minerals. 2nd ed. New York: Oxford University Press; 1997

[3] Środoń J. Nature of mixed-layer clays and mechanisms of their formation and alteration. Annual Review of Earth and Planetary Sciences. 1999;**27**(1):19-53

[4] Wilson MJ. Weathering of the Primary Rock-Forming Minerals: Processes. Products and Rates: Cambridge University Press; 2018

[5] Chamley H. Clay Sedimentology. Berlin: Springer-Verlag; 1989

[6] Dong H, Jaisi DP, Kim J, Zhang G. Microbe–clay mineral interactions. American Mineralogist. 2009;**94**(11-12):1505-1519

[7] Dong H et al. Mineral transformations associated with the microbial reduction of Fe(III) in nontronite. Chemical Geology. 2009;**264**(1-4):227-235

[8] Kostka JE, Dalton DD, Skelton H, Dollhopf S, Stucki JW. Growth of iron(III)-reducing bacteria on clay minerals as the sole electron acceptor and comparison of growth yields on a variety of oxidized iron forms. Applied and Environmental Microbiology. 2002;**68**(12):6256-6262

[9] Yang S, Liu D, Zheng W, Fan Q, Wang H, Zhao L. Microbial reduction and alteration of Fe (III)-containing smectites in the presence of biochar-derived dissolved organic matter. Applied Geochemistry. 2023;**152**:105661. DOI: 10.1016/j.apgeochem.2023.105661

[10] Lee K, Kostka JE, Stucki JW. Comparisons of structural Fe reduction in smectites by bacteria and dithionite: An infrared spectroscopic study. Clays and Clay Minerals. 2006;**54**:195-208. DOI: 10.1346/CCMN.2006.0540205

[11] Pentráková L, Su K, Pentrák M, Stucki JW. A review of microbial redox interactions with structural Fe in clay minerals. Clay Minerals. 2013;**48**(3):543-560

[12] Zhang G, Dong H, Kim J, Eberl DD. Microbial reduction of structural Fe^{3+} in nontronite by a thermophilic bacterium and its role in promoting the smectite-to-illite reaction. American Mineralogist. 2007;**92**(8-9):1411-1419

[13] Fan Q, Wang L, Fu Y, Li Q, Liu Y, Wang Z, et al. Iron redox cycling in layered clay minerals and its impact on contaminant dynamics: A review. Science of the Total Environment. 2023;**855**:159003. DOI: 10.1016/j.scitotenv.2022.159003

[14] Michalski JR, Cuadros J, Niles PB, Parnell J, Rogers AD, Wright SP. Groundwater activity on Mars and implications for a deep biosphere. Nature Geoscience. 2013;**6**(2):133-138

[15] Parenteau MN, Cady SL. Microbial biosignatures in iron-mineralized phototrophic mats at chocolate pots hot springs, Yellowstone National Park, United States. PALAIOS. 2010;**25**(2):97-111

[16] Lovley DR et al. Dissimilatory Fe(III) and Mn(IV) reduction. Microbiological Reviews. 1991;**55**(2):259-287

[17] Wilson MJ, editor. Clay Mineralogy: Spectroscopic and Chemical Determinative Methods. London: Springer Science and Business Media; 2012

[18] Broz AP. Organic matter preservation in ancient soils of earth and Mars. Life. 2020;**10**(7):113

[19] Posth NR, Canfield DE, Kappler A. Biogenic Fe(III) minerals: From formation to diagenesis and preservation in the rock record. Earth-Science Reviews. 2014;**135**:103-121

[20] Kostka JE, Wu J, Nealson KH, Stucki JW. The impact of structural Fe(III) reduction by bacteria on the surface chemistry of smectite clay minerals. Geochimica et Cosmochimica Acta. 1999;**63**:3705-3713

[21] Jung J, Yoo K-C, Rosenheim BE, et al. Microbial Fe(III) reduction as a potential iron source from Holocene sediments beneath Larsen ice shelf. Nature Communications. 2019;**10**:5786

[22] Vuillemin A, Morlock M, Paskin A, et al. Authigenic minerals reflect microbial control on pore waters in a ferruginous analogue. Geochemical Perspectives Letters. 2023;**28**:20-26

[23] Stucki JW. Microbial reduction of iron in smectite. Comptes Rendus Geoscience. 2006;**338**(6-7):468-475

[24] Bishop JL, Dobrea EZN, McKeown NK, Parente M, Ehlmann BL, Michalski JR, et al. Phyllosilicate diversity and past aqueous activity revealed at Mawrth Vallis, Mars. Science. 2008;**321**(5890):830-833

[25] Michalski JR, Niles PB. Deep crustal carbonate rocks exposed by meteor impact on Mars. Nature Geoscience. 2010;**3**(11):751-755

[26] Bristow TF, Rampe EB, Achilles CN, Blake DF, Chipera SJ, Craig P, et al. Clay mineral diversity and abundance in sedimentary rocks of Gale crater, Mars. Science Advances. 2018;**4**(6):eaar3330. DOI: 10.1126/sciadv.aar3330

[27] Rampe EB, Bristow TF, Archer PD, et al. Mineralogy of Vera Rubin ridge from the CheMin X-ray diffraction instrument on the curiosity rover on Mars. Journal of Geophysical Research: Planets. 2020;**125**(4):e2019JE006295

[28] Eigenbrode JL, Summons RE, Steele A, et al. Organic matter preserved in 3-billion-year-old mudstones at gale crater, Mars. Science. 2018;**360**(6393):1096-1101

[29] Carter J, Loizeau D, Mangold N, Poulet F, Bibring J-P. Widespread surface weathering on early Mars: A case for a warmer and wetter climate. Earth and Planetary Science Letters. 2015;**454**:49-58

[30] Vago JL, Westall F, Pasteur Payload Team, et al. Habitability on early Mars and the search for biosignatures with the ExoMars rover. Astrobiology. 2017;**17**(6-7):471-510

[31] Westall F, Foucher F, Bost N, et al. Biosignatures on Mars: What, where, and how? Implications for the search for Martian life. Astrobiology. 2015;**15**(11):998-1029

[32] Farmer JD, Des Marais DJ. Exploring for a record of ancient Martian life. Journal of Geophysical Research: Planets. 1999;**104**(E11):26977-26995

[33] Williford KH, Farley KA, Stack KM, Allwood AC, Beaty D, Beegle LW, et al. The NASA Mars 2020 rover mission and the search for extraterrestrial life. In: From Habitability to Life on Mars. Amsterdam: Elsevier; 2018. pp. 275-308

[34] Al-Muhaidib AI. Swelling characteristics of expansive soils in Riyadh, Saudi Arabia. Geotechnical and Geological Engineering. 2003;**21**(4):401-425

[35] Hudak PF. Geologic controls on foundation damage in north-central Texas. GeoJournal. 1998;**45**(3):59-164. DOI: 10.1023/A:1006965231244

[36] Nagaraj HB, Munnas MM, Sridharan A. Critical evaluation of determining swelling pressure by swell-load method and constant volume method. Geotechnical Testing Journal. 2009;**32**(4):305-314. DOI: 10.1520/GTJ102051

[37] Little DN, Nair S, Herbert B. Addressing Sulfate-Induced Heave in Lime-Treated Soils. Arlington, VA: National Lime Association; 2010

[38] Cornell University. Cation exchange capacity of soils. Cornell Cooperative Extension Agronomy Fact Sheet Series. 2007

[39] Fredlund DG, Sheng D, Zhao J. Estimation of soil suction from the soil-water characteristic curve. Canadian Geotechnical Journal. 2011;**48**(2):186-198

[40] Sahin H. Nondestructive Test Methods for Rapid Assessment of Flexible Base Performance in Transportation Infrastructures. 2014

[41] Yang H, Rahardjo H, Leong EC, Fredlund DG. Factors affecting drying and wetting soil-water characteristic curves of sandy soils. Canadian Geotechnical Journal. 2004;**41**(5):908-920

[42] Khoury N, Brooks R, Zaman MM, Khoury CN. Variations of resilient modulus of subgrade soils with postcompaction moisture contents. Transportation Research Record. 2009;**2101**(1):72-81. DOI: 10.3141/2101-09

[43] Hong Z, Bulut R, Aubeny CP, Jayatilaka R, Lytton RL. Prediction of moisture-induced soil heave in flexible pavements. Transportation Research Record: Journal of the Transportation Research Board. 2006;**1949**(1):148-156

[44] Houston WN, Dye HB, Zapata CE, Perera YY, Harraz A. Determination of SWCC using one-point suction measurement and standard curves. In: Unsaturated Soils 2006. 2006. pp. 1482-1493. DOI: 10.1061/40802(189)123

[45] Little DN, Nair S. Recommended Practice for Stabilization of Sulfate-Rich Subgrade Soils. Washington, DC: National Highway Cooperative Research Program, Transportation Research Board; 2009. Available from: https://scholar.google.com/scholar?cluster=8967075133502798500&hl=en&as_sdt=0,44

[46] Thompson D, Beasley DJ. SP-2209-OCN Handbook for Marine Geotechnical Engineering. Port Hueneme, CA: Naval Facilities Engineering Command; 2012. Available from: https://apps.dtic.mil/sti/html/tr/ADA556807/

[47] Basma AA, Al-Homoud AS, Husein A. Laboratory assessment of swelling pressure of expansive soils. Applied Clay Science. 1995;**9**(5):355-368. DOI: 10.1016/0169-1317(94)00032-L

[48] Bulut R. Finite element method analysis of slabs on elastic half space expansive soil foundations [Ph.D. dissertation]. College Station, TX: Texas A&M University; 2001. Available from: https://www.proquest.com/openview/6400066ec30550cb37c4012996e13ef6/1?pq-origsite=gscholar&cbl=18750&diss=y

[49] Germaine JT, Germaine AV. Geotechnical Laboratory Measurements for Engineers. Hoboken, NJ: Wiley; 2009

[50] Bell FG. Engineering Geology. 2nd ed. Amsterdam: Elsevier; 2007

[51] Bishop JL, Madejová J, Komadel P, Fröschl H. Remote detection of clay minerals. In: Bergaya F, Lagaly G, Beneke K, editors. Developments in Clay Science. Vol. 8. Amsterdam: Elsevier; 2017. pp. 381-412

[52] Van der Meer FD, van der Werff HMA, van Ruitenbeek FJA, Hecker CA, Bakker WH, Noomen MF, et al. Multi- and hyperspectral geologic remote sensing: A review. International Journal of Applied Earth Observation and Geoinformation. 2012;**14**(1):112-128

[53] Chabrillat S, Goetz AFH, Krosley L, Olsen HW. Use of hyperspectral images in the identification and mapping of expansive clay soils and the role of spatial resolution. Remote Sensing of Environment. 2002;**82**(2-3):431-445. DOI: 10.1016/S0034-4257(02)00060-3

[54] Belfiore CM, La Russa MF, Mazzoleni P, Pezzino A, Viccaro M. Technological study of "ghiara" mortars from the Historical City Centre of Catania (Eastern Sicily, Italy) and petro-chemical characterisation of raw materials. Environmental Earth Sciences. 2010;**61**(5):995-1003. DOI: 10.1007/s12665-009-0418-5

[55] Marchand THJ. The Djenné Mosque: World Heritage and Social Renewal in a West African Town. Amsterdam University Press; 2013. Available from: https://soas-repository.worktribe.com/output/440981

Chapter 2

Physical Purification of Egyptian Bentonite and Its Application as a Sustainable Binder in Iron Oxide Waste Recycling

Ahmed A. Khalifa and Eman M. Abdel Hamid

Abstract

The global demand for environmentally acceptable iron ore pelletization binders has increased the interest in natural alternatives, such as Egyptian bentonite. Despite its abundance and primary composition of montmorillonite, raw Egyptian bentonite has limited swelling capacity and poor industrial performance owing to excessive impurities and heterogeneous mineralogy. This chapter examines hydrocyclone separation as a sustainable and effective method for purifying Egyptian bentonite for pelletization. This chapter covers sodium and calcium bentonite clay varieties. After purification, bentonite clay is the most popular binder for iron ore pelletization because of its binding and mechanical qualities. This shaper examines the geological distribution and mineralogical features of Egypt's necessary bentonite deposits, focusing on montmorillonite-rich areas such as Ayun Musa and El-Barqan. Furthermore, this chapter covers advanced characterization methods (X-ray diffraction (XRD), X-ray Fluorescence (XRF), zeta potential) and complementary modifications. Bentonite is purified using physical processes, which involve mechanical separation, and chemical procedures, which utilize targeted reactions to change the clay's composition or surface chemistry. Chemical purification may damage the crystalline structure of montmorillonite; therefore, physical purification is safer and more effective for Egyptian bentonite. Hydrocyclone-based physical purification improves the purity of bentonite while conserving its mineralogical structure. Hydrocyclone-purified bentonite enhanced green and dried iron pellets. Purified bentonite improved the pellet hardening *via* thermal consolidation. The absence of toxic phases and sodium-rich smectite facilitated solid-state diffusion and liquid-phase formation. These variables created a strong sintering bridge. This chapter emphasizes the strategic necessity of valorizing local bentonite resources for sustainability and circular economy principles in the iron and steel industry.

Keywords: bentonite clay, El-Dekheila iron oxide waste, El-Barqan, purification, XRD, pelletization

1. Introduction

Bentonite is a naturally occurring clay primarily composed of smectite group minerals, among which montmorillonite constitutes the principal component. Montmorillonite, a key representative of the smectite family, typically originates from volcanic ash alteration. Depending on its geological origin, bentonite often contains various associated gangue minerals, such as quartz, feldspar, calcite, and gypsum. These impurities can adversely impact the clay's physicochemical properties, including swelling capacity, plasticity, and viscosity, limiting its suitability for specific industrial applications [1]. Consequently, beneficiation processes are commonly employed to enhance the quality and performance of bentonite for use in high-value applications [2].

Bentonite is commonly categorized into two main types—sodium bentonite (Na bentonite) and calcium bentonite (Ca bentonite)—based on the predominant exchangeable cation present [3, 4]. Sodium bentonite is generally preferred for industrial applications because of its superior swelling capacity, enhanced rheological behavior, and higher cation exchange capacity (CEC) than that of calcium bentonite. However, many naturally occurring bentonites, such as those found in Egypt, typically require beneficiation to meet the performance standards demanded by high-end applications, including drilling fluids, foundry molds, and iron ore pelletization [2, 5–8].

Among the diverse industrial applications of bentonite, its role as a binder in iron ore pelletization is particularly critical [7]. In this capacity, bentonite enhances the mechanical strength of the green pellets. This ensures the thermal stability of fired pellets, which are key properties influenced mainly by their swelling capacity and cation exchange potential. Bentonite's vital role in modern ore processing technologies is growing as the global iron and steel industry moves to lower-grade iron ores.

Due to demand for drilling, remediation of the environment, foundry, and iron ore pelletization, global bentonite production reached 21 million metric tons in 2024 [9, 10]. The leading producers worldwide are the United States, India, Turkey, China, and Greece. In the Arab and African regions, countries such as Egypt, South Africa, Morocco, Zimbabwe, and Algeria serve as key suppliers, owing to their substantial geological reserves and favorable mining conditions [11].

In Egypt, bentonite reserves in the northern part of the Western Desert hold considerable industrial potential. However, significant non-clay impurities, such as quartz, feldspars, and carbonates, often hinder their utilization. These negatively impact key properties like swelling capacity, limiting their suitability for advanced industrial applications [12]. To overcome these limitations, purification and beneficiation processes are essential to enhance the performance of the material. In a recent study [13], an integrated approach combining Landsat-derived land surface temperature (LST) data, principal component analysis (PCA), band ratio techniques, and XRD/XRF analyses was used to investigate clay deposits near Al-Alamein, Egypt. The study identified montmorillonite, kaolinite, and illite as the primary mineral constituents of the clay. A notable decline in the LST values of the exposed clay layers was observed, suggesting significant alterations in their physical and thermal characteristics. These findings underscore the value of integrating remote sensing technologies with mineralogical analyses to effectively assess and optimize local clay resources, such as Egyptian bentonite, for sustainable industrial applications, including its use as a binder in iron ore pelletization.

Several recent studies have demonstrated that physical beneficiation methods, especially hydrocyclone separation, can significantly enhance the quality of Egyptian bentonite. This is achieved by selectively removing coarse gangue minerals while concentrating on the smectite-rich fraction, thus improving the swelling index and cation exchange capacity [7, 12]. Compared to chemical or thermal treatments, hydrocycloning is a cost-effective, scalable, and environmentally sustainable alternative [7].

Conventional gravity separation techniques are generally ineffective for bentonite purification because of the similar specific gravities of clay and gangue minerals. However, wet classification methods, such as hydrocycloning, exploit differences in particle size rather than density [14, 15]. This method is particularly effective for sodium bentonite, whose particles disintegrate into fine fractions in water, thereby enabling efficient separation. In contrast, calcium bentonite tends to resist dispersion and may require grinding, which can complicate the separation process [12, 16].

A representative case study involving raw Ca bentonite from El-Barqan demonstrated the promise of hydrocyclone technology for upgrading local bentonite. This study explored the effects of operational parameters, such as vortex diameter, apex size, and feed pressure, on performance indicators, including particle size distribution, swelling index, and cation exchange capacity. These findings support the potential of hydrocyclone separation as a viable method for enhancing the industrial applicability of Egyptian bentonite [7].

Despite abundant international research on bentonite processing, a notable gap remains in the literature concerning the practical purification challenges specific to Egyptian bentonite deposits. Existing studies often overlook the unique mineralogical characteristics of local clays and provide limited insight into the scalability and industrial relevance of physical separation methods, such as hydrocycloning, particularly with regard to their effectiveness as binders in iron ore pelletization.

This review seeks to address this gap by offering a comprehensive evaluation of the physical beneficiation of El-Barqan bentonite using hydrocyclone technology. This study emphasizes the potential of this method to enhance the quality of local bentonite, making it a viable, sustainable, and efficient binder for iron ore pelletization and recycling of steelmaking by-products, particularly in the agglomeration of iron oxide residues generated by the El-Dekheila steel plant.

1.1 Background and industrial relevance of bentonite

Bentonite is widely regarded as an essential industrial clay due to its unique physicochemical properties, including high adsorption capacity, swelling behavior, thixotropic gel formation, and cation exchange capabilities. These characteristics make bentonite suitable for over 40 industrial applications spanning multiple sectors, such as drilling fluids [17], wastewater treatment and environmental remediation [18], pharmaceuticals and medicine [19, 20], construction, ceramics, paints, and agriculture [21].

In the iron and steel industries, bentonite is conventionally used as a binder for producing iron ore pellets, imparting green strength, shape integrity, and thermal resistance to the pellets during firing. However, associated gangue minerals, such as quartz, feldspar, calcite, and mica, can adversely affect its industrial performance, reduce the montmorillonite content, and introduce impurities that degrade product quality and process efficiency [2]. Consequently, the beneficiation or purification of bentonite is often necessary to enhance its suitability for specific industrial applications, particularly in high-performance sectors such as metallurgy and environmental engineering.

1.2 Crystal structure and chemical properties of bentonite

Bentonite is a type of clay mineral that belongs to the family of layered silicates, characterized by a 2:1 phyllosilicate structure. This structure consists of a central octahedral sheet, typically composed of aluminum or magnesium, sandwiched between two outer tetrahedral sheets rich in silica. This mica-type configuration is fundamental to the physical and chemical behaviors of these minerals. As illustrated in **Figure 1**, variations in the composition and occupancy of the octahedral and tetrahedral layers among different clay minerals result in distinct structural and functional properties of the clay minerals. Bentonite, which is primarily composed of the mineral montmorillonite, is classified as an expanding clay due to its ability to absorb water and undergo significant swelling, in contrast to non-expanding clays such as illite [22, 23].

Clay minerals are silicate sheets that combine tetrahedral and octahedral layers and are typically less than 2 mm thick. The tetrahedral sheets comprise SiO_4 units sharing three oxygen atoms in a hexagonal pattern, whereas the fourth (apical) oxygen projects toward the adjacent octahedral sheet. The octahedral layers typically contain cations such as Al, Mg, and Fe, which are coordinated with six oxygen or hydroxyl anions in an octahedral geometry [22]. Bentonite predominantly comprises montmorillonite, a mineral phase of the smectite group. The layered structure of montmorillonite allows for the reversible incorporation of water molecules and exchangeable cations (such as Na^+ or Ca^{2+}) within the interlayer spaces [24]. This unique structural flexibility gives rise to essential properties, such as swelling capacity, hydration, and high cation exchange capacity (CEC), making bentonite highly effective in various industrial applications.

Bentonite is generally categorized into three main types based on the dominant exchangeable cation in its interlayer structure: sodium (Na^+), calcium (Ca^{2+}), or a mixed Na-Ca form [14, 25]. Its high swelling capacity and dispersibility distinguish sodium bentonite (**Figure 2**) [26], making it particularly suitable for applications such as drilling muds, sealing materials, and iron ore pelletization [27]. Calcium bentonite, on the other hand, displays lower swelling and plasticity; however, it can be chemically modified using sodium carbonate (Na_2CO_3) to mimic the behavior of sodium bentonite. Therefore, the industrial performance and economic viability of bentonite are closely linked to its montmorillonite content, structural integrity, and nature of its interlayer cations.

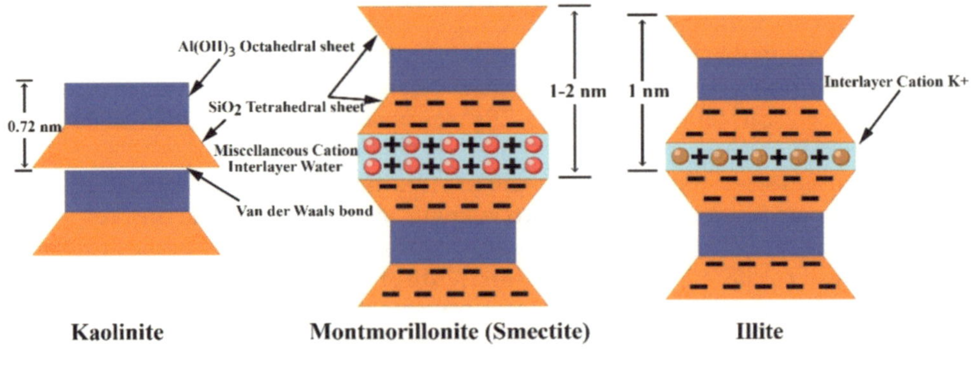

Figure 1.
Structural classification of different clay minerals.

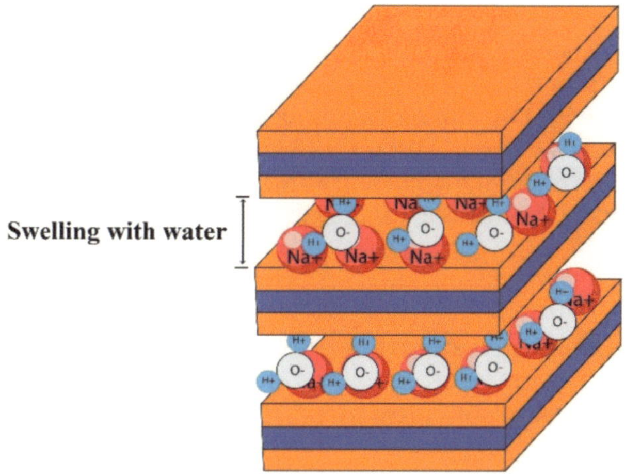

Figure 2.
Swelling of montmorillonite by hydration of the interlayer.

1.3 Types of bentonite

Two main types of bentonite clays are generally recognized based on the predominant exchangeable cations in the interlayer spaces: sodium (Na) and calcium (Ca) bentonite. Sodium bentonite, which is dominated by Na^+ ions, exhibits a high affinity for water and can absorb large quantities of water, resulting in significant swelling. In contrast, calcium bentonite, characterized by the prevalence of Ca^{2+} ions, shows limited water absorption and is therefore classified as a low-swelling bentonite [12, 27]. A study by Muhammad and Siddiqua [27] corroborates this, indicating that the cation exchange capacity (CEC) of calcium bentonite is 50 meq/100 g, which is inferior to that of high swelling-type bentonite (Na), which ranges from 80 to 85 meq/100 g. Additional distinctions encompass the surface area of Ca bentonite clay, ranging from 60 to 120 m^2/g, whereas Na bentonite clay measures approximately 20–30 m^2/g [28].

The chemical compositions of bentonites also vary between the two types. XRF analyses revealed that Ca bentonite clay generally contains higher CaO and MgO contents, whereas Na bentonite is enriched in Na_2O. For instance, representative Ca bentonite samples consist of CaO (4.88%), MgO (2.65%), Fe_2O_3 (6.35%), K_2O (0.69%), Na_2O (0.74%), Al_2O_3 (16.44%), SiO_2 (60.82%), and LOI (6.79%) [29]. In comparison, Na bentonite comprises CaO (1.44%), MgO (2.75%), Fe_2O_3 (11.92%), K_2O (0.29%), Na_2O (3.22%), Al_2O_3 (15.34%), SiO_2 (52.55%), TiO_2 (1.62%), and LOI (9.80%) [30].

X-ray diffraction (XRD) studies further confirmed the structural characteristics of both types. For Ca bentonite, montmorillonite is the predominant clay mineral, with accessory minerals such as quartz, albite, and clinoptilolite [27]. Natural Na bentonite, on the other hand, is also rich in montmorillonite, with characteristic reflections at d001 = 14.29 Å and d020 = 4.49 Å, confirming its sodium-rich interlayer composition [31]. Minor peaks of quartz and feldspar are typically observed.

Having distinguished bentonite into sodium and calcium types based on their exchangeable cations, physicochemical behavior, and mineralogical features, it is crucial to contextualize these classifications within the Egyptian context. The geological formation of bentonite, typically through the alteration of volcanic ash, is strongly linked to its locality, which explains the global variation in bentonite typessuch as, sodium bentonite deposits in Wyoming and Japan and calcium bentonite deposits exploited elsewhere for their cost-effectiveness and environmental benefits. Mixed bentonites, characterized by minimal swelling and the presence of accessory minerals, further influence the industrial value of bentonite deposits. Accordingly, the following section addresses the occurrence and characterization of bentonite deposits in Egypt, highlighting their geological distribution, formation processes, and compositional variability.

2. Occurrence and characterization of bentonite deposits in Egypt

2.1 Geological distribution and formation

Bentonitic clays in Egypt are predominantly located in the northern part of the Western Desert and are interbedded with sedimentary rocks such as siltstone, sandstone, and marly limestone. These deposits vary in thickness from less than 1 meter to 10 meters and commonly exhibit faint lamination, although more massive beds or lenticular forms are also observed. Clays typically display a yellowish to greenish-gray coloration with a waxy luster and are frequently coated with iron oxyhydroxides along structural surfaces.

Agha et al. reported that the first documented occurrence of bentonite in Egypt was near El Fayoum City [30]. Since then, several commercial deposits have been identified in four main provinces of the northern Western Desert, with stratigraphic ages ranging from the Middle Eocene to the Late Pliocene [12]. Although some of these deposits exhibit high kaolinite content, the origin of all occurrences, particularly their association with volcanic precursors, remains a subject of ongoing research. Mineralogical investigations have revealed a broad compositional spectrum, including trioctahedral Ca-Mg-smectite, Na-montmorillonite, kaolinite, and illite-smectite mixed layers, along with non-clay minerals such as quartz, feldspar, calcite, and gypsum [12, 29].

The genesis of these bentonitic clays is linked to complex geological processes that occurred during several key geological intervals, particularly the Upper Eocene, Lower Miocene, and Middle Miocene. These conditions promote the formation of Fe-rich smectite and kaolinite varieties, with smectite typically being the dominant phase [12].

2.2 Regional distribution

Bentonite deposits in Egypt occur across diverse stratigraphic and tectonic settings, ranging from the Paleozoic to the Cenozoic eras. Major occurrences are found in the Nile Valley, the Delta region, and the Eastern and Western Deserts. These deposits exhibit significant mineralogical and chemical variability, which directly affects their purity and industrial suitability. **Table 1** summarizes the principal Egyptian bentonite deposits, their geological context, dominant minerals, and potential applications [12, 30].

Bentonite clay deposits are distributed across several localities in Egypt. **Figure 3** illustrates the geographical distribution of significant bentonite ore deposits in the country. These deposits exhibit varying degrees of mineralogical purity, alteration processes, and suitability for industrial use. Notably, bentonite deposits in the

Location	Geological age	Main minerals	Industrial applications
El-Barqan (Western Desert)	Jurassic-Cretaceous	Montmorillonite, kaolinite, quartz	Water purification, environmental liners
South El-Hammam (Eastern Desert)	Lower Miocene (Moghra formation)	Montmorillonite, kaolinite, albite, calcite	Drilling muds, ceramics, and swelling-based products
Fayoum Region	Miocene	Smectite, kaolinite, quartz	Lightweight aggregates, ceramics, and soil conditioning
Alamein Area (North Coast)	Cenozoic sediments	Fe-rich smectite, kaolinite	Ceramic production, rheological additives
Wadi El-Natrun	Undifferentiated	Kaolinite, feldspar, quartz	Refractory materials, industrial fillers
Red Sea Coast (Halaib Region)	Meso-Cenozoic	Smectite, illite, kaolinite, feldspar	Plastic industries, swelling agents, and geotechnical barriers

Table 1.
Summary of major Egyptian bentonite deposits and their characteristics.

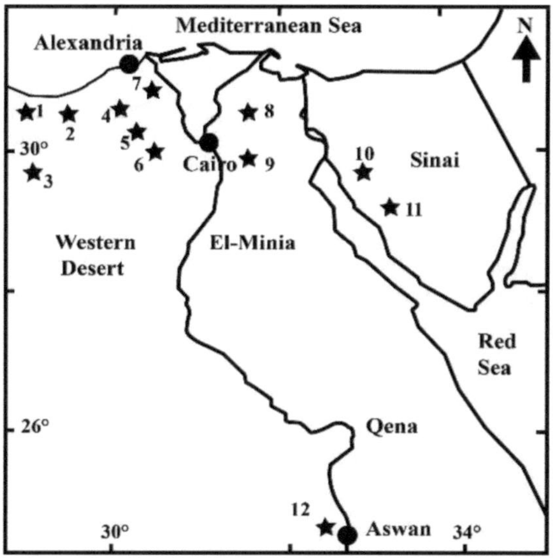

Figure 3.
Occurrences of bentonite clay in Egypt. (1) Alamein deposits; (2) El-Hammam deposits; (3) Qattara Depression deposits; (4) Wadi El-Natrun deposits; (5) Qasr El-Sagh deposits; (6) Kom Aushim deposits; (7) Cairo-Alexandria road deposits; (8) Hamza deposit; (9) Qattami deposits; (10) Ayun Musa deposits; (11) Abu Zeneima deposits; (12) Kurkur deposits.

northern Western Desert, particularly those at El-Barqan, demonstrate superior compositional quality and are currently the focus of most commercial exploitation efforts in Egypt [32].

These mineralogical and spatial variations highlight the importance of region-specific assessments for sustainable resource exploitation in the area. Understanding the geological origin and mineral composition, developing tailored processing strategies that maximize industrial value and minimize environmental impact [12, 13].

2.3 Mineralogical and statistical studies

Among Egyptian bentonite deposits, the northern Western Desert is the only region currently under active commercial exploitation, highlighting its superior industrial potential [32]. A detailed mineralogical and industrial assessment of 12 quarries across four provinces [30] revealed significant variability in mineral composition, impurity levels, and industrial suitability, which directly affects the achievable purity after physical beneficiation.

In the southwestern Alamein area (P1), the bentonitic clays predominantly comprise iron-rich smectite and coarsely crystalline kaolinites. These mineralogical attributes confer desirable rheological and thermal properties, rendering the clay suitable for applications in drilling muds and ceramic products.

The El-Barqan area, located south of El-Hammam (P2), is characterized by a higher proportion of iron-poor illite and quartz, which contributes to its elevated chemical purity. This makes the deposit particularly well-suited for construction materials and environmental liners, where low iron content is a critical criterion.

In the Wadi El-Natrun area near Deir El Baramous (P3), the clays were primarily composed of finely crystalline kaolinite accompanied by feldspar, indicating a distinct genesis compared to other regions. These mineralogical features suggest potential use in refractory applications and as industrial fillers, especially in applications where high thermal stability is essential.

The El Fayoum Depression (P4) hosts smectite-rich clays with minor illite, reflecting their high swelling capacity and water-retention ability. These properties support their applicability as drilling fluids and soil conditioners in agricultural contexts. Agha et al. [30] applied discriminant function analysis (DFA) to the mineralogical dataset of Egyptian bentonite deposits, demonstrating a high classification accuracy of 92%. This statistical validation confirmed the mineralogical heterogeneity and geological complexity of the studied regions. Integrating advanced techniques such as X-ray diffraction (XRD) with multivariate statistical tools such as DFA enabled the precise classification of bentonite types, thereby supporting targeted exploitation and sustainable resource utilization. These findings underscore the industrial significance and resource potential of the northern Western Desert, reaffirming its position as Egypt's premier bentonite production region.

The chemical and physical properties of bentonite vary owing to the nature and concentration of montmorillonite, the primary mineral component, and the type and quantity of associated secondary minerals. Consequently, the chemical composition of bentonite may differ not only between distinct deposits but also within the same quarry site. It is important to note that the values reported in **Table 2** represent raw, naturally occurring Egyptian bentonite, as documented in previous studies, and are indicative of regional deposits.

This mineralogical variability directly affects purification efficiency and industrial suitability. Alamein and Fayoum clays contain higher levels of iron-bearing or swelling minerals that complicate beneficiation, while Wadi El-Natrun clays are dominated by feldspar and kaolinite, which limit their binder suitability. In contrast, the El-Barqan deposit is characterized by relatively low iron-bearing phases, making it particularly advantageous for iron ore pelletization. Purification of this deposit is therefore focused primarily on the removal of free quartz, without the additional requirement of iron-removal steps that are necessary for other deposits. In this context, hydrocyclone technology has gained increasing attention as a physical purification method, particularly effective for El-Barqan bentonite, where it can efficiently

Location	Application	SiO_2	Al_2O_3	Fe_2O_3	CaO	MgO	Na_2O	Reference
El-Barkan	Binder	52.20	18.40	8.40	1.16	1.70	0.33	[7]
Abu Zenima, South Sinai	Bleaching of oils and drilling mud	42.40	11.62	7.06	10.50	1.89	1.59	[2]
North Qaret Shamekh	—	50.62	21.77	9.83	2.50	2.09	2.09	[32]
Deir El Gamel	—	50.11	19.64	7.89	2.13	1.02	1.43	[32]
Qur El Malik	—	53.23	22.71	9.85	0.82	2.52	2.73	[32]
Average of West and Southwest El Qattara Depression bentonite	—	51.04	20.41	8.52	1.32	1.97	2.14	[32]
South of El-Hammam City	Pharmaceutical applications	51.08	16.37	9.27	0.97	2.68	0.84	[33]

Table 2.
Chemical compositions of various Egyptian bentonites have been reported in previous studies.

remove deleterious impurities such as quartz and carbonates, thereby enhancing its performance as a binder in pelletization.

3. Bentonite as a binder in iron ore pelletization

Fine iron ore powder cannot be directly utilized in blast furnaces or most direct reduction furnaces because of its poor flowability and low permeability, which hinder efficient gas–solid interactions. Consequently, it must be agglomerated through pelletization to produce a feedstock with the requisite mechanical strength and metallurgical properties for effective processing. Therefore, pellet production cannot be achieved without using binders to agglomerate the iron ore concentrate.

An ideal pellet binder should possess the following characteristics:

- It can be easily distributed throughout the pellet
- Facilitate the movement and retention of water within the pellet
- Strong interparticle bonds were created within the pellets

Bentonite effectively fulfills these requirements because of its high swelling capacity, excellent cation exchange ability, and notable rheological plasticity. These properties enable it to disperse uniformly within the pellet matrix, retain moisture efficiently, and enhance particle cohesion during the pellet growth.

The most commonly used inorganic binder, bentonite, is notable for having this effect. Bentonite clays help regulate the water presented on the pellet's surface, enabling a wider range of effective pellet moistures and allowing pelletization at workable growth rates. The most common inorganic bentonite binder is silica-containing clay, which can contribute to slag binding during the sintering process. However, this same silica can lead to a reduction in the overall reducibility because it is largely impervious to reducing gases. Similarly, there are concerns that colemanite could harm reducibility because it promotes slag fusion at lower firing temperatures. Perhaps the one universal requirement of pellet binders is that they must be able to reach every part of the pellet that is to be bound: a binder that does not have some form of mobility within the material will not succeed at binding the pellet. For example, mechanically increasing the dispersion of bentonite clays significantly improves their effectiveness [34]. Many binders can contain undesirable impurities (especially free silica impurities, such as quartz, which are often associated with bentonite) and are expensive; therefore, it is usually preferred to use less binder where possible.

Bentonite clay is the most commonly used binder in iron ore pelletization due to its favorable binding and mechanical properties [35, 36]. However, conventional bentonite contains up to 85% silica and alumina, which are considered undesirable acidic gangue materials in ironmaking. These impurities can adversely affect the quality of the final iron product by increasing the slag volume and reducing the furnace efficiency.

To address these challenges, considerable research has focused on reducing the amount of bentonite used in pellet formulations. One promising strategy involves the use of purified bentonite, free from deleterious gangue minerals and other contaminants. This approach has been shown to enhance the physical integrity and chemical purity of the resulting pellets, thereby improving their performance in ironmaking processes.

3.1 Role of bentonite in pelletization

In the iron ore pelletization process, moist iron ore particles and flux materials are agglomerated into spherical pellets with the assistance of a binder (**Figure 4**). These green pellets are initially held together by surface tension and capillary forces; however, they tend to disintegrate upon drying without incorporating an appropriate binder. Bentonite, the most commonly used binder, plays a crucial role throughout the pelletization process, from green ball formation to the final thermal hardening stage [37].

During green ball formation, bentonite enhances interparticle cohesion by regulating capillary and viscous forces, thereby promoting structural integrity during wet agglomeration. It also controls moisture transport within the green pellet matrix by increasing the internal viscosity, which slows down water migration and stabilizes the pellet growth rates. Additionally, bentonite mitigates green ball spalling during drying by enhancing thermal shock resistance, increasing the drying temperature, and improving energy efficiency.

During the induration phase, bentonite localizes at particle contact points, forming mechanical bridges that reinforce the structure of the dry pellets. As a low-melting-point silicate, bentonite undergoes partial melting upon heating, generating a liquid phase that facilitates particle bonding, promotes the sintering of iron oxide grains, and enhances pellet densification.

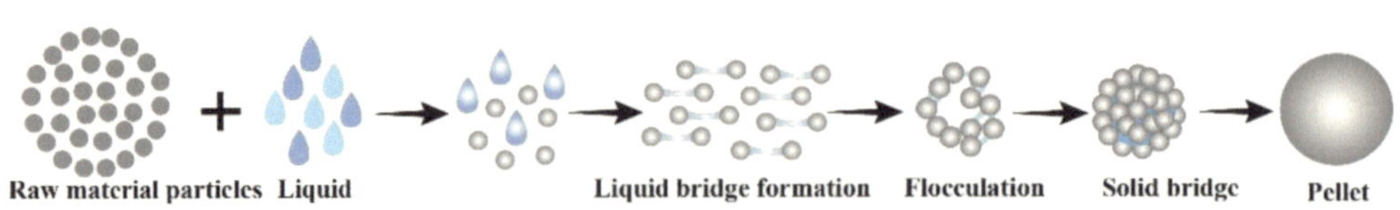

Figure 4.
Schematic of green ball growth in iron ore in iron ore pelletization.

DOI: http://dx.doi.org/10.5772/intechopen.1012867

Despite ongoing research and the growing adoption of organic binders, such as polyacrylamide, in iron ore pelletization, bentonite continues to dominate this industry. Approximately 90% of iron ore pellets produced in the United States utilize bentonite-based binders, whereas only approximately 10% are manufactured using polyacrylamide as an organic alternative. This continued reliance on bentonite is primarily attributed to the insufficient mechanical strength of organic binders. Consequently, bentonite remains essential for ensuring the durability of the handling characteristics of green pellets during processing and transportation [38].

3.2 Desirable pellet properties

Iron ore pellets must meet stringent physical, chemical, and metallurgical requirements to be suitable for use in blast furnaces and direct reduction processes. The desirable attributes include:

- High mechanical strength to withstand handling and transportation.
- Narrow size distribution (typically 9–12 mm) to ensure optimal bed permeability and efficient gas-solid contact.
- High reducibility, minimizing the time and energy required for reduction reactions.
- Low gangue content increases Fe grade and minimizes slag formation.

Green pellet strength is critical during the drying and transport stages and is typically evaluated using drop number, compressive strength, and thermal shock resistance tests. After induration, fired pellets must resist abrasion and retain their structural integrity during shipping and handling processes.

From a metallurgical perspective, pellets must exhibit high reducibility, low reduction disintegration index (RDI), and controlled swelling to avoid fine generation and maintain furnace efficiency. In direct reduction processes, such as MIDREX, the Fe content should exceed 64%, and the pellet basicity should be optimized to influence the microstructure and improve the reduction behavior.

The selection and quality of the binder used during green pellet formation are critical for achieving the desired characteristics of iron ore pellets, particularly mechanical strength and thermal stability. Bentonite remains the most commonly employed among the various binders available due to its cost-effectiveness and high swelling capacity. However, the performance of bentonite is primarily determined by its montmorillonite content, which directly influences its binding strength, pellet porosity, and reducibility.

In this context, the present study investigated the physical purification of El-Barqan bentonite using hydrocyclone technology. The objective was to enhance the clay's purity and improve its performance as a sustainable binder in iron ore pelletization processes.

4. Purification of bentonite

The industrial value of bentonite is closely linked to its mineralogical composition, particularly the montmorillonite content, which governs its swelling, adsorption, and binding capabilities [31]. Enhancing the purity of bentonite can significantly expand its industrial applications and improve its performance in metallurgical processes.

The associated gangue compounds include gravel, quartz, feldspar, calcite, and iron oxides [39]. However, Egyptian bentonite typically exhibits a polycationic nature and commonly occurs in mixed sodium-calcium forms. It often contains other clay minerals, such as illite and kaolinite, in addition to substantial amounts of non-clay impurities [11]. Bentonite is a naturally occurring substance; however, it is never pure. It usually contains impurities such as quartz and cristobalite (silica, SiO_2), alumina (Al_2O_3), feldspars, carbonates, zeolites, iron oxides, and organic materials.

These contaminants reduce the effectiveness of the iron pelletization process [40]. These acidic constituents counteract prior beneficiation efforts and necessitate the addition of more acidic fluxing agents to attain the desired basicity within reduction furnaces, including blast furnaces and direct reduction iron (DRI) units [40]. This, in turn, increases the slag volume and elevates energy consumption, as a significant portion of energy is wasted in melting the excess slag [41]. Therefore, bentonite purification is critical to improve its consistency and suitability for industrial use, particularly as a sustainable binder in iron ore pelletization [14]. Two principal approaches are commonly employed to purify bentonite: physical methods, which rely on mechanical separation techniques, and chemical methods, which aim to remove impurities through targeted reactions that alter the composition or surface chemistry of the clay. Physical purification typically involves techniques such as sieving, magnetic separation [42], sedimentation [43], ultrasonic dispersion, and centrifugation [44] to isolate the <2 μm montmorillonite fraction and remove associated impurities such as quartz and kaolinite. Among these, centrifugation at 8000 rpm for 40 minutes has been reported as the most effective method [28, 39]. These approaches are cost-effective, simple, and environmentally friendly, making them suitable for large-scale applications [45].

In contrast, chemical purification focuses on removing soluble salts, carbonates, organic matter, and oxides of Fe, Al, Si, Ti, and Mn. Silica and alumina can be extracted using a 5% sodium carbonate solution, whereas amorphous Fe and Al oxides are typically removed using oxalate solutions or acid treatments. Organic matter is typically eliminated using hydrogen peroxide, sodium hypochlorite, or disodium peroxodisulfate. Chemical purification methods, including acid activation and ion exchange, are employed to enhance specific properties, such as surface area, cation exchange capacity, and adsorption performance [43, 46]. However, these methods often involve high costs, hazardous reagents, and additional steps to eliminate residual chemicals. Chemical purification methods may suffer from a fundamental drawback: the potential dissolution of montmorillonite (Mt) during treatment, which can damage its crystalline structure. Veiskarami et al. confirmed this issue [47], reporting that applying a chemical purification method removed most associated impurities; however, it was accompanied by a slight decrease in the d001 value, indicating partial destruction of the montmorillonite structure. The choice between these methods depends on the targeted application, impurity type, and sustainability requirements.

In this context, chemical purification should be approached with caution because of its potential effect on the crystalline structure of montmorillonite. Therefore, several studies have investigated the use of physical purification methods on Egyptian bentonite as a safer and more effective alternative.

4.1 Physical purification of Egyptian bentonite

M.S. Hassan et al. [2] conducted an extensive investigation on bentonite samples from the Nukhul Formation in the Abu Zenima region of Sinai. The study aimed to evaluate the suitability of these samples for industrial uses such as drilling mud and oil bleaching. The raw bentonite was subjected to a multistage beneficiation process comprising crushing, scrubbing, hydrocyclone separation, acid leaching, and alkaline and acid activation.

X-ray diffraction (XRD) analysis (**Figure 5**) confirmed that montmorillonite remained the dominant clay mineral after purification, while the intensities of quartz and kaolinite peaks were significantly reduced. Moreover, the peaks corresponding to calcite and dolomite were eliminated, indicating the effective removal of these gangue minerals. The enhanced intensity of the smectite peak, along with improved chemical composition and thermal behavior (as evidenced by thermogravimetric analysis), pointed to a substantial enrichment of the clay fraction. These findings suggest hydrocyclone-based physical purification can enhance bentonite purity while preserving its essential mineralogical structure.

Although purification of Egyptian bentonite has been widely explored for several applications, including drilling fluids, pharmaceuticals, and agriculture, the utilization of hydrocyclone technology remains relatively under-investigated, particularly in the context of bentonite use as a binder in iron ore pelletizing. This gap is especially relevant for bentonite deposits with challenging mineralogical compositions, such as those in the El-Barqan region, which are typically characterized by low montmorillonite content and high concentrations of associated gangue minerals, including silica, gypsum, and iron oxides such as hematite. In this context, El-Barqan bentonite has been highlighted in recent research due to its mineralogical complexity and industrial relevance. The application of hydrocyclones as compact, energy-efficient devices

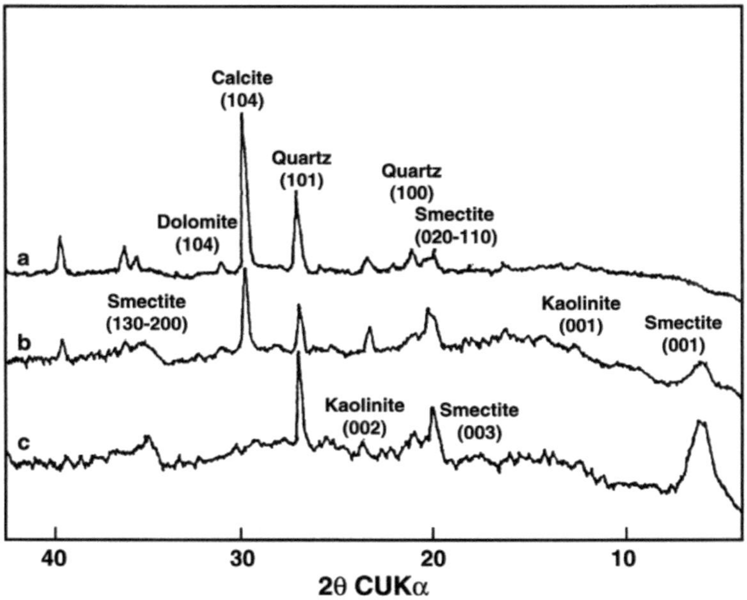

Figure 5.
XRD for (a) raw bentonite, (b) purified bentonite, and (c) activated bentonite [2].

capable of separating fine particles based on density and size has been proposed as a practical and sustainable approach for the physical purification of bentonite resources. Unlike traditional mechanical classifiers or centrifuges, hydrocyclones operate without moving parts and require minimal maintenance, which makes them especially well-suited for large-scale industrial clay processing [48–51]. Given their growing significance in clay beneficiation, the following section presents a detailed discussion of their working principles, separation mechanisms, and key design parameters.

4.2 Hydrocyclones: Applications and design advances

Hydrocyclones play a critical role in particle classification and dewatering during mineral processing. Specifically, they enhance the quality of bentonite by removing coarse impurities, such as sand and silt, thereby improving the clay's purity and functional performance.

The separation mechanism within a hydrocyclone is governed by centrifugal sedimentation. The feed slurry enters tangentially, generating a high-velocity rotational flow that forms two distinct vortices:

- An outer, downward-moving vortex that transports coarser particles to the underflow region.

- An inner, upward-moving vortex that carries finer particles to overflow.

The efficiency of this separation process is highly dependent on the geometric design and operational parameters of the hydrocyclone. Key design features influencing internal flow dynamics and separation efficiency include the formation of an air core, recirculation zones, and the locus of zero vertical velocity (LZVV). Operational parameters, such as feed pressure, solids concentration, and particle size distribution, also play pivotal roles. For instance, increasing inlet velocity enhances centrifugal force and improves separation efficiency, albeit at the cost of higher energy consumption [52, 53]. Conversely, elevated solid concentrations and slurry viscosity can hinder flow dynamics and reduce classification sharpness [54, 55].

The geometry of the hydrocyclone, particularly the cone angle, vortex finder, and inlet configuration, significantly affects its performance. Recent advancements in cone design, including parabolic and hyperbolic geometries, have demonstrated improved separation sharpness and efficiency [56, 57].

Hydrocyclone performance is typically assessed using a partition curve, which represents the probability of a particle of a given size reporting to the underflow. Two critical performance indicators were derived from this curve:

- Cut size (d_{50}): The particle size at which there is a 50% probability of reporting to the underflow or overflow.

- Sharpness of separation: A measure of the system's ability to accurately distinguish between fine and coarse particles [58].

A comprehensive understanding of these indicators is essential for optimizing the design and operation of hydrocyclones. Ongoing research continues to refine these aspects using advanced hydrodynamic modeling and design optimization techniques [59]. Achieving effective hydrocyclone performance in bentonite purification requires a holistic integration of fluid dynamics knowledge, geometric optimization, and accurate operational control.

4.3 Operational variables affecting hydrocyclone performance

Building upon the discussion of design aspects, the performance of a hydrocyclone is also significantly governed by key operational parameters such as feed flow rate, fluid properties, and the characteristics of suspended solids [58].

4.3.1 Feed flow rate and pressure drop

The feed velocity is critical for enhancing the centrifugal forces within the hydrocyclone, thereby improving particle separation and underflow recovery [56]. For instance, a twofold increase in feed velocity can nearly double the tangential velocity, resulting in a 56% improvement in recovery. However, excessively high velocities may induce turbulence and short-circuiting, compromising separation efficiency [60, 61]. Pressure drop is another key factor, as it directly influences axial and tangential velocities. An increased pressure drop reduces the cut size and enhances separation precision [62].

4.3.2 Density and rheology of the suspension

The density of the suspension affects the migration behavior of the particles. Optimal separation is achieved when the medium density lies between the densities of the particle fractions [63]. Also, high viscosity or pseudoplastic rheological behavior can dampen centrifugal forces, reducing separation efficiency [54, 64].

4.3.3 Solids concentration and particle characteristics

Increasing solids concentration generally leads to higher energy losses and a reduction in tangential velocity, resulting in decreased separation efficiency and a larger cut size [65]. High solids content can also destabilize the flow field and reduce the sharpness of separation [66]. Furthermore, excessive solids may cause spigot blockages, leading to significant performance degradation [54, 67].

Maintaining optimal operational conditions and carefully controlling feed characteristics are essential for maximizing hydrocyclone efficiency, particularly in applications such as bentonite purification.

4.4 Hydrocyclone separation

The hydrocyclone test rig employed in this study is illustrated in **Figure 6**. It consisted of a 25 mm hydrocyclone mounted vertically above a 50 L feed tank. A 5 HP centrifugal pump delivered the slurry to the hydrocyclone feed inlet, while a bypass line equipped with a control valve was installed on the outlet to regulate the desired pressure drop. A pressure gauge was positioned near the feed inlet to monitor the operating pressure throughout the experiments.

Both design and operating parameters of the hydrocyclone were systematically examined. The design parameters and experimental operating conditions are summarized in **Table 3**.

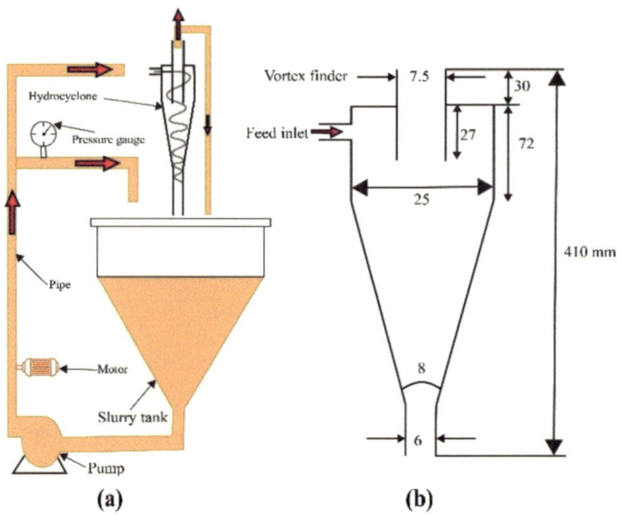

Figure 6.
(a) Hydrocyclone rig (b) hydrocyclone dimension.

Parameter	Symbol	Dimension (mm)
Diameter of the hydrocyclone	Dc	25
Cylindrical section length	H	72
Feed inlet diameter	Di	7.5
Vortex finder diameter	Do	7.5
Underflow diameter	Du	6
Vortex finder length (inner)	I	8
Cone angle	θ	8°
Feed pressure	P	20, 30, 60 psi
Feed solid concentration	Cs	5, 10, 15, 20%

Table 3.
Hydrocyclone design parameters and operating parameters.

To demonstrate the practical relevance of hydrocyclone technology in clay beneficiation, the El-Barqan region in Egypt's Northern Western Desert was selected as a case study due to its industrial importance and abundant bentonite reserves. However, geological surveys have revealed that large sections of these deposits contain low-grade bentonite, highlighting the necessity of purification before industrial use. Accordingly, a detailed beneficiation study [7] employed hydrocyclone separation to upgrade the bentonite's quality, focusing on its application as a binder in iron ore pelletization.

The beneficiation procedure involved grinding the raw clay to below 45 μm, followed by classification using a 25 mm Mozley hydrocyclone. Under the optimal operating conditions of 5 wt.% feed solids and 60 psi feed pressure, the overflow product was substantially enriched in montmorillonite (increasing from 29.5–51%), while quartz decreased significantly, and gibbsite was eliminated. Chemically, SiO_2 dropped from 52.20 to 42.35%, accompanied by an increase in Al_2O_3 and other clay-related oxides, confirming the efficiency of the beneficiation process. As illustrated in **Figure 7**, the particle size distribution of the feed, overflow, and underflow further validates the separation performance, with d_{50} values of approximately 6 μm for the overflow and 14 μm for the underflow. The kaolinite fraction remained nearly stable, showing only a slight decrease from 19.2 to 17.9%, while hematite increased modestly from 6.8 to 7.4%, most likely due to its association with the finer particles concentrated in the overflow.

These mineralogical and chemical changes are further corroborated by the flowchart (**Figure 8**) and XRD analysis (**Figure 9**), which reveal intensified montmorillonite peaks, reduced quartz, and the complete removal of gibbsite. Complementary XRF analysis (**Table 4**) confirms a significant reduction in SiO_2 alongside an enrichment in Al_2O_3 and other clay-forming oxides, highlighting the effectiveness of hydrocyclone purification in producing high-grade bentonite

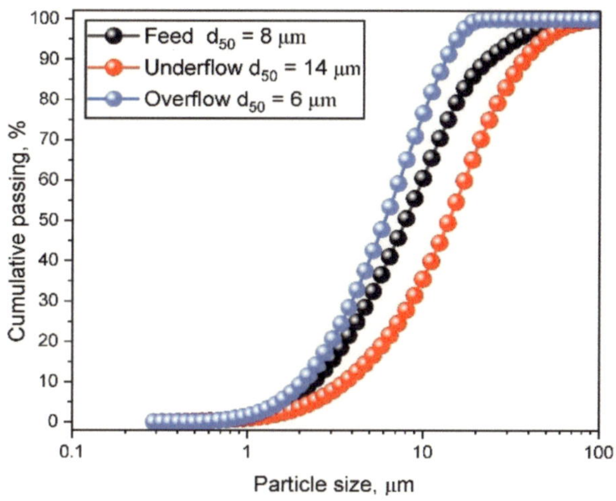

Figure 7.
Particle size distribution curves of the feed, overflow, and underflow obtained at the optimum hydrocyclone operating conditions.

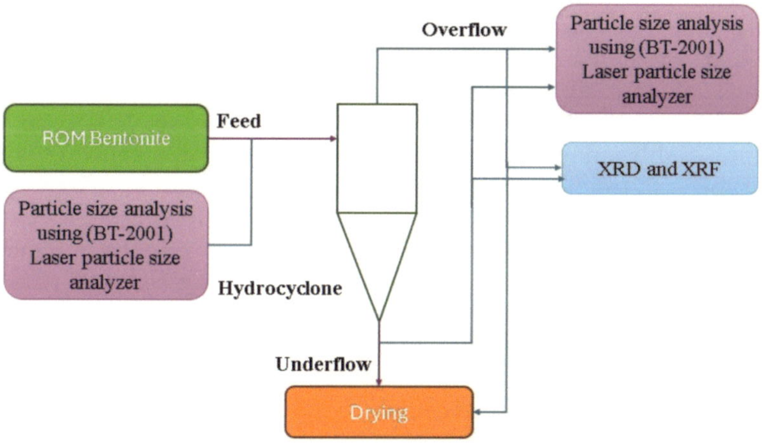

Figure 8.
Schematic flowchart of El-Barqan bentonite beneficiation via hydrocyclone classification.

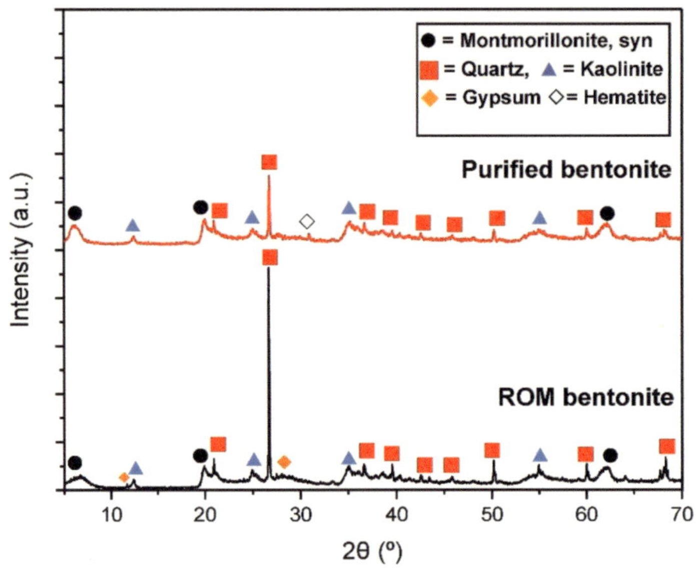

Figure 9.
XRD patterns of raw and hydrocyclone-treated El-Barqan bentonite samples.

Oxides content	SiO_2	Al_2O_3	Fe_2O_3	CaO	MgO	Na_2O	K_2O	TiO_2	L.O.I
Purified El-Barqan	42.35	19.75	12.4	1.89	1.76	0.4	0.86	1.62	15.8
Wyoming bentonite	62.79	20.96	3.81	0.01	2.44	2.84	0.06	0.14	6.74

Table 4.
Chemical analysis of El-Barqan bentonite after hydrocyclone beneficiation [2].

suitable for sustainable use as a binder in iron ore pelletization. A comparative chemical analysis further illustrates these differences: Purified Wyoming bentonite is characterized by higher silica ($SiO_2 \approx 62.8\%$) and sodium ($Na_2O \approx 2.8\%$) contents together with relatively low iron oxides ($Fe_2O_3 \approx 3.8\%$), reflecting its sodium-dominated montmorillonite composition known for superior swelling and cation exchange capacity [68]. In contrast, El-Barqan Egyptian bentonite, even after hydrocyclone beneficiation, exhibits lower silica ($\approx 42.3\%$) and sodium ($\approx 0.4\%$) but substantially higher iron oxides ($\approx 12.4\%$) along with calcium and potassium enrichment. This composition suggests that while El-Barqan bentonite contains appreciable amounts of montmorillonite, it is more representative of a Ca-K-rich type with inherently lower swelling potential and exchange capacity compared to Wyoming sodium bentonite. Nonetheless, its properties could be significantly improved through Na-exchange treatment, thereby enhancing its suitability for industrial applications [33, 69].

Further insight was provided by zeta potential measurements (**Figure 10**), which showed that purified bentonite exhibited more negative surface charges across the pH range compared with the raw sample. This shift is attributed to the increased montmorillonite fraction, reduced quartz content, and exposure of more active surface sites. Enhanced surface negativity improves cation adsorption, swelling behavior, and electrostatic interaction properties that directly contribute to stronger pellet bonding and higher green and fired pellet strength.

Overall, hydrocyclone classification under optimized conditions effectively upgraded El-Barqan bentonite, demonstrating its potential as a sustainable and high-performance binder for iron ore pelletization.

The study evaluated the purified bentonite's functional performance in iron ore pellet formation based on these improved mineralogical and chemical characteristics. The following section details the raw materials used and the experimental procedures adopted to assess the pellet quality and validate the industrial applicability of the upgraded clay.

5. Practices of pelletization at the laboratory scale

Controlling green pellet production during pelletization has been a significant challenge on a laboratory scale due to its critical impact on the quality of final pellets [70]. These challenges primarily stem from variations in raw material properties, moisture content inaccuracies, and the need to optimize pelletizing parameters. Several efforts have been undertaken to standardize laboratory pelletization procedures [34]. Pelletization was carried out using a laboratory-scale disc pelletizer operated at a tilt angle of 60°, a rotation speed of 17 rpm, and a residence time of 10 minutes. The experimental procedure involved the following steps:

1. *Ore preparation:* The purified bentonite was mixed with El-Dekheila iron oxide waste (particle size <75 μm) to serve as the feed material.

2. *Seed formation:* An initial 80 g of the prepared mixture was used to form seed pellets. Water was gradually added while stirring to facilitate the agglomeration of fine iron oxide particles.

3. *Balling stage:* The remaining feed mixture and additional water were added alternately to the pelletizing disc to promote pellet growth. Once most pellets reached a diameter between 8 and 18 mm, water addition—limited to 10 wt.%—was stopped. Pellets were then rolled for an additional 2 minutes to ensure densification. A total of 400 g of feed material was used per batch.

4. *Green pellets:* Qualified pellets with a diameter between 10 and 13.2 mm were separated and dried at 110°C for 8 hours before mechanical strength testing.

5. *Final pellets (induration):* The thermal treatment simulated industrial induration and evaluated the influence of binder quality on final pellet strength. The heat cycle included drying at 200°C, preheating at 1000°C, and firing at 1250–1300°C under oxidizing conditions. The temperature was increased at 10°C/min, with 15-minute holding periods at both the preheating and firing stages [7].

5.1 Mechanical testing of pellets

Mechanical characterization of the green, dried, and indurated pellets was conducted to evaluate their strength and durability under standard testing conditions. A sufficient number of pellets were tested in parallel for each performance parameter, and average values were calculated to ensure data reliability and minimize measurement variability.

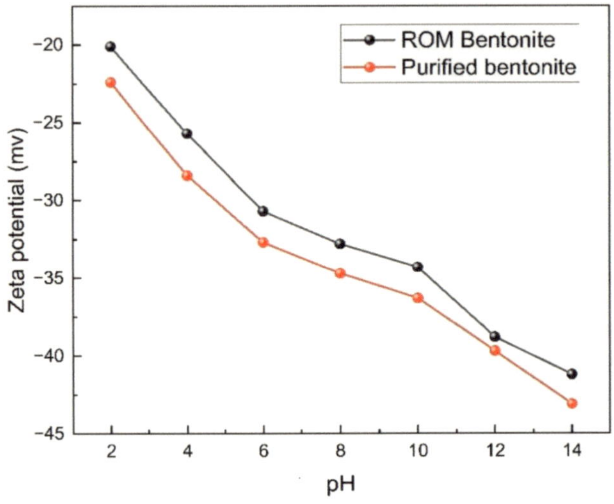

Figure 10.
Variation of zeta potential with pH for raw and purified bentonite samples.

The testing procedures involved the following methods:

1. Drop number test:

   Ten green and dried pellets were individually dropped from a height of 46 cm onto a 10 mm-thick steel plate. The number of drops required to fracture each pellet was recorded, and the average value was used to indicate impact resistance.

2. Compressive strength of green pellets:

   A calibrated mechanical press was used to apply a vertical load of up to 5 kg on each green pellet until failure occurred. The average failure load was used to assess the structural integrity of the green pellets.

3. Compressive strength of indurated pellets:

   After undergoing the induration process, which involved staged thermal treatment at 200°C (drying), 1000°C (preheating), and 1250°C (firing) under oxidizing conditions, the fired pellets were tested for compressive strength using a MEGA.KSC-10 hydraulic press. The heating rate was maintained at 10°C/min, with holding times of 15 minutes at both the preheating and firing stages.

   The results of these mechanical evaluations highlight the significant role of induration conditions and binder properties in determining the final mechanical performance of the pellets, which will be examined in detail in the following section.

5.2 Purified bentonite as a binder in pelletization

Based on laboratory-scale experiments, Khalifa et al. [7] assessed the effectiveness of hydrocyclone-purified bentonite from the El-Barqan region as a binder in pelletizing finely ground El-Dekheila iron oxide waste (<75 μm).

The findings revealed that increasing bentonite content from 0.5 to 1.5 wt.% significantly improved the quality of both green and dried pellets. Green pellet drop strength rose from 2.5 to 3.5 drops, while dried pellets reached a drop strength of 7 drops at 1.5 wt.% binder content. Compressive strength exhibited a parallel improvement. These enhancements were attributed to the higher montmorillonite concentration and reduced quartz content in the purified bentonite, which improved water absorption, swelling behavior, and binding efficiency. Moreover, the finer particle size (d_{90} = 17.33 μm, d_{50} = 7.54 μm) facilitated uniform dispersion and stronger interparticle bonding.

These results underscore the critical role of montmorillonite's layered structure in forming water bridges during pellet formation, which solidify into strong bonds upon drying [71–74]. Further densification and cohesion occurred due to structural contraction during drying. Thus, hydrocyclone-treated bentonite proved to be a high-performance binder, offering a practical route to valorizing low-grade clay for sustainable pelletization applications and reducing reliance on imported materials.

5.3 Induration treatment

Following the initial drying of green pellets at 110°C, mechanical testing revealed a progressive improvement in drop number and compressive strength with increasing binder content (up to 1.5 wt.%). This enhancement was attributed to removing quartz impurities and enriching montmorillonite, which promoted stronger interparticle bonding. The layered structure and swelling capacity of montmorillonite facilitated the formation of water bridges during drying, which subsequently solidified into durable bonds. These results agree with earlier studies [75], highlighting the key role of montmorillonite in improving green pellet integrity.

Upon subsequent firing, purified bentonite contributed to more effective pellet hardening by enhancing thermal consolidation. The absence of deleterious phases and sodium-rich smectite improved solid-state diffusion and liquid-phase formation conditions. These factors collectively led to the development of robust sintering bridges. In conclusion, the results presented in this chapter demonstrate that hydrocyclone-purified bentonite significantly enhances both green and fired pellet properties. These improvements establish its potential as a sustainable, technically effective, and economically viable binder for the recycling of industrial iron oxide waste [27].

5.4 Final pellet strength

Further analysis of the sintered pellets confirmed that the mechanical strength improvements were closely associated with the chemical and mineralogical transformations induced by the purified bentonite. Increasing the binder dosage up to 1.5 wt.% led to a notable rise in compressive strength at a sintering temperature of 1250°C, reflecting the superior bonding efficiency of the purified clay.

This enhancement was primarily linked to the partial melting of bentonite during firing, which facilitated the formation of a low-melting eutectic slag that promoted effective sintering through solid interparticle bridges and the crystallization of dissolved phases. These microstructural developments contributed to reduced porosity and improved internal cohesion, which agrees with previous findings by Mohamed et al. [27]. While the authors observed improved pellet strength despite a decrease in the basicity ratio (CaO/SiO_2) from 0.46 to 0.28 due to increased silica content from the purified bentonite, the influence of basicity was not thoroughly investigated. The added bentonite also contributes to elevated levels of acidic oxides such as Al_2O_3 and SiO_2, which can significantly alter the slag chemistry and liquid-phase formation during sintering.

However, the study did not account for the cumulative effect of these acidic oxides on the pellet's basicity. It did not consider corrective measures such as adjusting flux additions (e.g., CaO or MgO) to counterbalance the increased acidity. Since basicity is critical in determining slag viscosity, sintering temperature range, and phase formation, a more comprehensive analysis is necessary to ensure optimal pellet consolidation conditions.

Thus, it is recommended that future studies incorporate a detailed assessment of the basicity ratio and its modulation through additive control, especially when using high-purity or acid-rich binders. This would enable a clearer understanding of the interplay between binder composition, melt phase evolution, and mechanical performance of the sintered product.

6. Conclusions and future perspectives

This chapter provides a comprehensive evaluation of the physical purification of El-Barqan bentonite using a laboratory-scale hydrocyclone, with particular emphasis on enhancing its suitability as an industrial binder in iron ore pelletization. The results demonstrate that hydrocycloning effectively upgrades locally sourced, low-grade bentonite by reducing non-clay gangue minerals such as quartz and enriching the montmorillonite fraction. The purification process led to marked improvements in the bentonite's physicochemical properties, including swelling capacity, cation exchange potential, and colloidal stability. Mineralogical and chemical analyses confirmed increased alumina and montmorillonite contents alongside a reduction in free silica, while zeta potential and particle size measurements indicated enhanced surface reactivity and fineness, both of which are critical for uniform dispersion and strong interparticle adhesion during pellet formation.

Hydrocyclone operating parameters—namely feed pressure, vortex finder diameter, and pulp concentration—were identified as critical factors governing purification efficiency, underscoring hydrocycloning as a scalable and eco-efficient beneficiation technique aligned with modern green metallurgy practices. Experimental pelletization trials using the purified bentonite demonstrated notable improvements in the physical integrity of both green and dry pellets, as reflected by higher drop numbers and compressive strength. Following induration at 1250°C, the pellets exhibited enhanced mechanical properties resulting from the formation of silicate bonding phases and a cohesive glassy matrix, which promoted densification, the development of sintering bridges, and improved thermal stability.

Looking forward, several research and industrial directions emerge. Pilot- and industrial-scale trials are required to validate laboratory findings and assess the long-term performance of purified bentonite in continuous pelletizing operations. Integration of hydrocyclone purification with complementary physical or chemical upgrading techniques—such as selective flocculation, mild thermal activation, or organo-modification—could further enhance bentonite functionality for applications beyond ironmaking, including environmental remediation, drilling fluids, and catalysis. Additionally, techno-economic and life-cycle assessments are recommended to evaluate the cost-effectiveness and environmental impact of large-scale hydrocycloning relative to conventional beneficiation methods.

Ultimately, the results support the proposition that purified El-Barqan bentonite can be a viable domestic alternative to imported high-grade bentonite binders. This contributes to improved self-sufficiency, promotes the circular utilization of ironmaking by-products such as those generated by the El-Dekheila steel plant, and opens new avenues for the local valorization of Egypt's bentonite resources. The chapter thus underscores the broader applicability of physical beneficiation strategies in enhancing the industrial value of underutilized natural clays in resource-constrained regions.

Author details

Ahmed A. Khalifa[1] and Eman M. Abdel Hamid[2*]

1 Department of Minerals Beneficiation and Agglomeration, Central Metallurgical Research and Development Institute (CMRDI), El Tebbin, Cairo, Egypt

2 Department of Chemical Engineering, Egyptian Academy for Engineering and Advanced Technology (EAEAT), Cairo, Egypt

*Address all correspondence to: engahmedkhalifa2@gmail.com

References

[1] Temraz MG, Hassanien I. Mineralogy and rheological properties of some Egyptian bentonite for drilling fluids. Journal of Natural Gas Science and Engineering. 2016;**31**:791-799. DOI: 10.1016/j.jngse.2016.03.072

[2] Hassan MS, Abdel-Khalek NA. Beneficiation and applications of an Egyptian bentonite. Applied Clay Science. 1998;**13**(2):99-115. DOI: 10.1016/S0169-1317(98)00021-0

[3] Murray HH. Chapter 6 Bentonite Applications. In Developments in Clay Science. Amsterdam, Netherlands: Elsevier B.V. 2006;2:111-130. DOI: 10.1016/S1572-4352(06)02006-X

[4] Shah LA, Khattak NS, Valenzuela MGS, Manan A, Valenzuela Díaz FR. Preparation and characterization of purified Na-activated bentonite from Karak (Pakistan) for pharmaceutical use. Clay Minerals. 2013;**48**(4):595-603. DOI: 10.1180/claymin.2013.048.4.03

[5] Abdou MI, Al-sabagh AM, Dardir MM. Evaluation of Egyptian bentonite and nano-bentonite as drilling mud. Egyptian Journal of Petroleum. 2013;**22**(1):53-59. DOI: 10.1016/j.ejpe.2012.07.002

[6] El-Mahllawy MS, Sharara AM, Hassaan MM, Abdel Haleem AM. The composition and activation aspects of El-Fayoum clays for using as a drilling fluid. Egyptian Journal of Petroleum. 2013;**22**(3):395-404. DOI: 10.1016/j.ejpe.2013.10.009

[7] Khalifa AA, Hassan ESRE, Elbendari AM, Amin R, Abdel-Khalek MA. Physical purification of Egyptian bentonite clay as a sustainable binder of El-Dekheila-iron-oxide-waste. Euro-Mediterranean Journal for Environmental Integration. 2025;**10**(4):2761-2772. DOI: 10.1007/S41207-025-00773-7/METRICS

[8] Hasan K, Jaya RP, Yahaya FM. Application of bentonite in cement-based composites: A review of current status, challenges and future prospects. Journal of Building Engineering. 2024;**98**:111171. DOI: 10.1016/J.JOBE.2024.111171

[9] Zhuang G, Li Q, Zhang Z, Bergaya F, Yuan P. Chapter 2 - Application of bentonite in water-based drilling fluids. In: Clay Science in Drilling and Drilling Fluids: Developments in Clay Science. Amsterdam, Netherlands: Elsevier B.V. 2024;11:21-53. DOI: 10.1016/B978-0-443-15598-7.00005-5

[10] Akkaya U, Çokça E. Use of waste foundry sand, bentonite, and waste rubber as a landfill cap layer material. Innovative Infrastructure Solutions. 2023;**8**(5):1-8. DOI: 10.1007/s41062-023-01121-w

[11] U.S.G. Survey. Mineral Commodity Summaries. VA, USA: Reston, 2025. DOI: 10.3133/MCS2025

[12] Agha M, Ferrell RE, Hart GF. Mineralogy of Egyptian bentonitic clays I: Discriminant function analysis. Clays and Clay Minerals. 2012;**60**(4):387-404. DOI: 10.1346/CCMN.2012.0600405

[13] Agha MA, Masoud AA, Madani AA, Abdel Aziz SM. Clarify the relationship between the land surface temperature and clay minerals using Landsat temporal data, spectral, and XRD data: A case study from some clay quarries southwest Al-Alamin City, Egypt. Kuwait Journal of Science. 2025;**52**(2):100371. DOI: 10.1016/J.KJS.2025.100371

[14] Boylu F, Çinku K, Esenli F, Çelik MS. The separation efficiency of Na-bentonite by hydrocyclone and characterization of hydrocyclone products. International Journal of Mineral Processing. 2010;**94**(3-4):196-202. DOI: 10.1016/j.minpro.2009.12.004

[15] Bloodworth AJ, Morgan DJ, Briggs DA. Laboratory processing trials on kaolin-bearing sandstones from Pugu, Tanzania, using conventional and new hydrocyclone bodies. Clay Minerals. 1989;**24**(3):539-548. DOI: 10.1180/claymin.1989.024.3.07

[16] Schneider M, Neeße T. Overflow-control system for a hydrocyclone battery. International Journal of Mineral Processing. 2004;74:S339-S343. DOI: 10.1016/j.minpro.2004.07.037

[17] Afolabi RO, Orodu OD, Efeovbokhan VE. Properties and Application of Nigerian Bentonite Clay Deposits for Drilling Mud Formulation: Recent Advances and Future Prospects. Applied Clay Science. 2017;**143**:39-49. DOI: 10.1016/j.clay.2017.03.009

[18] Shayesteh A, Fard MY, Vafaeian M. Assessment of one and two-dimensional molecular diffusion of chloride ion through ceramic powder waste-bentonite mixture as landfill liner. Construction and Building Materials. 2022;**341**:1-18. DOI: 10.1016/j.conbuildmat.2022.127826

[19] Elshater A, Elhaddad A, Elattaar A, Abugharbia M, Soliman W. Characterisation of the Egyptian Pliocene bentonite from the Sohag region for pharmaceutical use. Arabian Journal of Geosciences. 2018;**11**(14):1-12. DOI: 10.1007/s12517-018-3691-3

[20] Shah LA, Valenzuela, MGS, Ehsan, AM, Valenzuela Díaz, FR, Khattak NS. Characterization of Pakistani purified bentonite suitable for possible pharmaceutical application. Applied Clay Science. 2013;**83-84**:50-55. DOI: 10.1016/j.clay.2013.08.007

[21] Sameh HI, Mabrouk MS, Ali AA, Elwalead AA. Synthesis and characterization of bentonite nanocomposites from Egyptian bentonite clay. International Journal of Nanotechnology and Allied Sciences. 2017;**1**(1):16-29

[22] Parolo ME, Savini MC, Vallés JM, Baschini MT, Avena MJ. Tetracycline adsorption on montmorillonite: pH and ionic strength effects. Applied Clay Science. 2008;**40**(1-4). DOI: 10.1016/j.clay.2007.08.003

[23] Kumari N, Mohan C. Basics of clay minerals and their characteristic properties. Clay and Clay Minerals. 2021:1-29. DOI: 10.5772/intechopen.97672

[24] Faghihian H, Godazandeha N. Synthesis of nano crystalline zeolite Y from bentonite. Journal of Porous Materials. 2009;**16**(3):331-335. DOI: 10.1007/s10934-008-9204-0

[25] Bain JA, Morgan DJ. Laboratory separation of clays by hydrocycloning. Clay Minerals. 1983;**18**(1). DOI: 10.1180/claymin.1983.018.1.04

[26] Endo M, Sato H. Swelling stress of bentonite: Thermodynamics of interlayer water in K-montmorillonite in consideration of alteration. Minerals. 2024;**14**(4):1-16. DOI: 10.3390/MIN14040430

[27] Mohamed OA, Shalabi MEH, El-Hussiny NA, Khedr MH, Mostafa F. The role of normal and activated bentonite on the pelletization of barite iron ore concentrate and the quality of pellets. Powder Technology. 2003;130:277-282. DOI: 10.1016/S0032-5910(02)00205-X

[28] Thuc CNH, Grillet AC, Reinert L, Ohashi F, Thuc HH, Duclaux L. Separation and purification of montmorillonite and polyethylene oxide modified montmorillonite from Vietnamese bentonites. Applied Clay Science. 2010;**49**(3):229-238. DOI: 10.1016/j.clay.2010.05.011

[29] El-Shabiny AM, Hammad SM, Ibrahim IA, Ismail AK. Combined thermal and structural study of some Egyptian bentonitic clays. Journal of Thermal Analysis. 1996;**46**(5):1421-1435. DOI: 10.1007/BF01979255

[30] Agha MA, Ferrell RE, Hart GF, El Ghar MSA, Abdel-Motelib A. Mineralogy of egyptian bentonitic clays II: Geologic origin. Clays and Clay Minerals. 2014;**61**(6):551-565. DOI: 10.1346/CCMN.2013.0610608

[31] Deng L, Lu Y, Shang X, Liu Z, Dong Y. Study on wet purification of Keerjian bentonite. Journal of Physics: Conference Series. 2023;2463:1-5. DOI: 10.1088/1742-6596/2463/1/012041

[32] Hamimi Z, El-Barkooky A, Frías MJ, Fritz H, Abd El-Rahman Y. The Geology of Egypt. Switzerland: Springer Nature Switzerland AG; 2020. doi: org/10.1007/978-3-030-15265-9

[33] Abdel-Motelib A, Kader ZA, Ragab YA, Mosalamy M. Suitability of a Miocene bentonite from North Western Desert of Egypt for pharmaceutical use. Applied Clay Science. 2011;**52**(1-2):140-144. DOI: 10.1016/j.clay.2011.02.012

[34] Kawatra SK, Claremboux V. Iron ore pelletization: Part I. Fundamentals. Mineral Processing and Extractive Metallurgy Review. 2022;43(4):529544. DOI: 10.1080/08827508.2021.1897586

[35] De Moraes SL, Kawatra SK. Laboratory study of an organic binder for peptization of a magnetite concentrate. Minerals and Metallurgical Processing. 2010;**27**(3):148-151. DOI: 10.1007/BF03402238/METRICS

[36] de Moraes SL, de Lima JRB, Ribeiro TR, de Moraes SL, de Lima JRB, Ribeiro TR. Iron ore pelletizing process: An overview. Iron Ores and Iron Oxide Materials. 2018:41-59. DOI: 10.5772/INTECHOPEN.73164

[37] Alsaqoor S, Borowski G, Alahmer A, Beithou N. Using of adhesives and binders for agglomeration of particle waste resources. Advances in Science and Technology. Research Journal. 2022;**16**(3):124-135. DOI: 10.12913/22998624/149456

[38] Halt JA, Kawatra SK. Review of Organic Binders for Iron Ore Concentrate Agglomeration. Minerals and Metallurgical Processing, 2014;31(2):73-94. DOI: 10.1007/ bf03402417

[39] Chipera SJ, Bish DL. Baseline studies of the clay minerals society source clays: Powder X-ray diffraction analyses. Clays and Clay Minerals. 2001;**49**(5):398-409. DOI: 10.1346/CCMN.2001.0490507

[40] Ngara T, Mavengere S, Bright S, Mapamba L. Investigating the effectiveness of organic binders as an alternative to bentonite in the pelletization of low grade iron ore. Physicochemical Problems of Mineral Processing. 2023;**59**(6):1-12. DOI: 10.37190/PPMP/176094

[41] Kawatra SK. New paradigms for iron ore pelletization. In: Davis B, et al. Extraction. The Minerals, Metals & Materials Series. Cham: Springer; 2018. pp. 953-959. DOI: 10.1007/978-3-319-95022-8_76

[42] Chipera SJ, Guthrie GD, Bish DL. Preparation and purification of mineral

dusts. Reviews in Mineralogy and Geochemistry. 1993;**28**(1):235-249

[43] Shah LA, das da Silva Valenzuela MG, Farooq M, Khattak SA, Díaz FRV. Influence of preparation methods on textural properties of purified bentonite. Applied Clay Science. 2018;**162**:155-164. DOI: 10.1016/j.clay.2018.06.001

[44] Bahranowski K et al. Influence of purification method of Na-montmorillonite on textural properties of clay mineral composites with TiO2 nanoparticles. Applied Clay Science. 2017;**140**:75-80. DOI: 10.1016/j.clay.2017.01.032

[45] Ciftci H. An introduction to montmorillonite purification. In: Montmorillonite Clay. London, UK: IntechOpen; 2021. DOI: 10.5772/intechopen.98188

[46] Kolta GA, Novak I, El-Tawil SZ, El-Barawy KA. Evaluation of bleaching capacity of acid-leached egyptian bentonites. Journal of Applied Chemistry and Biotechnology. 1976;**26**(7):355-360. DOI: 10.1002/jctb.5020260152

[47] Veiskarami M, Sarvi MN, Mokhtari AR. Influence of the purity of montmorillonite on its surface modification with an alkyl-ammonium salt. Applied Clay Science. 2016;**120**:111-120. DOI: 10.1016/j.clay.2015.10.026

[48] Saengchan K, Nopharatana A, Songkasiri W. New hydrocyclone design for high starch yield: Effect of cylindrical-conical length proportion on the starch, sulfur and protein separation efficiencies. Powder Technology. 2019;**346**:1-8. DOI: 10.1016/j.powtec.2019.01.077

[49] Bicalho IC, Mognon JL, Shimoyama J, Ataíde CH, Duarte CR. Separation of yeast from alcoholic fermentation in small hydrocyclones. Separation and Purification Technology. 2012;**87**:62-70. DOI: 10.1016/j.seppur.2011.11.023

[50] Banerjee C, Cepeda E, Switzer D, Hunter S. Application potential of Cavex® double effect hydrocyclone for the classification of mine tailings–a pilot scale study. Mineral Processing and Extractive Metallurgy Review. 2023;**45**(8):858-868. DOI: 10.1080/08827508.2023.2298376

[51] Lv W et al. Process of removing heavy metal ions and solids suspended in micro-scale intensified by hydrocyclone. Journal of Cleaner Production. 2020;**263**:1-12. DOI: 10.1016/j.jclepro.2020.121533

[52] Chakraborti N, Miller JD. Fluid flow in hydrocyclones: A critical review. Mineral Processing and Extractive Metallurgy Review. 1992;**11**(4):211-244. DOI: 10.1080/08827509208914207

[53] Jena SK, Sahoo H, Rath SS, Rao DS, Das SK, Das B. Characterization and processing of iron ore slimes for recovery of iron values. Mineral Processing and Extractive Metallurgy Review. 2015;**36**(3):174-182. DOI: 10.1080/08827508.2014.898300

[54] Gonçalves SM, Ullmann G, Morimoto MG, de Souza Barrozo MA, Vieira LGM. Effect of rheology and solids concentration on hydrocyclones performance: A study involving the design variables of an optimized hydrocyclone. Journal of Petroleum Science and Engineering. 2022;**210**:1-13. DOI: 10.1016/j.petrol.2021.110019

[55] Alves DG, da Silva JTT, Quintino DB, Ataíde CH. Desander mini-hydrocyclones applied to the separation of microspheres and sand

in non-Newtonian fluid: Efficiencies and drag analysis. Separation and Purification Technology. 2020;**234**:1-9. DOI: 10.1016/j.seppur.2019.116131

[56] Jiang L, Liu P, Yang X, Zhang Y, Li F. Comparative classification studies of red mud by using hydrocyclones. Minerals Engineering. 2019;**131**:124-130. DOI: 10.1016/j.mineng.2018.11.012

[57] Liu B et al. Internal reuse of methanol-to-olefin wastewater based on micro-channel separation coupling hydrocyclone regeneration. Journal of Environmental Management. 2023;**345**:1-11. DOI: 10.1016/j.jenvman.2023.118791

[58] de Sa da Silva DP, Morimoto MG, de Oliveira Silva D, Barrozo MAS, Vieira LGM. Effect of operational and geometric variables in Hydrocyclones to improve separation efficiency and reduce energy consumption: A review and new insights. Mineral Processing and Extractive Metallurgy Review. 2024:1-32. DOI: 10.1080/08827508.2024.2418636

[59] Zhao Z, Zhou L, Liu B, Cao W. Computational fluid dynamics and experimental investigation of inlet flow rate effects on separation performance of desanding hydrocyclone. Powder Technology. 2022;**402**:1-9. DOI: 10.1016/j.powtec.2022.117363

[60] Xiong Z, Xu J, Liu C. Interaction effects of inlet velocity and apex diameter on the separation performance of two-stage cone hydrocyclones. Powder Technology. 2023;**422**:1-14. DOI: 10.1016/j.powtec.2023.118446

[61] Zhang X et al. A natural monoterpene enol for dispersive liquid–liquid microextraction based on solidification of floating organic droplets for determination of benzophenone compounds in water samples. Separations. 2023;**10**(1):1-10. DOI: 10.3390/separations10010001

[62] Rocha CAO, Ullmann G, Silva DO, Vieira LGM. Effect of changes in the feed duct on hydrocyclone performance. Powder Technology. 2020;**374**:283-289. DOI: 10.1016/j.powtec.2020.07.001

[63] Chowdhury R, Billah MM, Roy S, Banik SC, Barua A. Analyzing the effect of the density of medium on efficiency of hydrocyclone separator in sorting of PVC and PET using CFD. Results in Materials. 2024;**21**:1-10. DOI: 10.1016/j.rinma.2023.100497

[64] Morimoto MG, Gonçalves SM, Barrozo MA d S, Vieira LGM. Study on filtering Hydrocyclones operating with concentrated non-Newtonian suspensions. Industrial and Engineering Chemistry Research. 2023;**62**(37):15190-15201. DOI: 10.1021/acs.iecr.3c02342

[65] Lager T, Forssberg KSE. Current processing technology for antimony-bearing ores a review, part 2. Minerals Engineering. 1989;**2**(4):543-556. DOI: 10.1016/0892-6875(89)90088-5

[66] Li F, Liu P, Yang X, Zhang Y, Zhao Y. Effects of inlet concentration on the hydrocyclone separation performance with different inlet velocity. Powder Technology. 2020;**375**:337-351. DOI: 10.1016/j.powtec.2020.07.110

[67] Gonçalves SM, Barrozo MAS, Vieira LGM. Effects of solids concentration and underflow diameter on the performance of a newly designed Hydrocyclone. Chemical Engineering and Technology. 2017;**40**(10):1750-1757. DOI: 10.1002/ceat.201600496

[68] Kiviranta L, Kumpulainen S, Quality Control and Characterization of Bentonite Materials; (Report No.

POSIVA-WR-11-84). Helsinki, Finland: Posiva Oy; 2011. 102 p

[69] Dardir FM, Mohamed AS, Abukhadra MR, Ahmed EA, Soliman MF. Cosmetic and pharmaceutical qualifications of Egyptian bentonite and its suitability as drug carrier for Praziquantel drug. European Journal of Pharmaceutical Sciences. 2018;**115**:320-329. DOI: 10.1016/j.ejps.2018.01.041

[70] Li C, Bai Y, Ren R, Liu G, Zhao J. Study of the mechanism for improving green pellet performance with compound binders. Physicochemical Problems of Mineral Processing. 2019;**55**(1):153-162. DOI: 10.5277/ppmp18116

[71] Eisele TC, Kawatra SK. A Review of Binders in Iron Ore Pelletization. Abingdon, UK: Taylor & Francis; 2003;24(1):1-90. DOI: 10.1080/08827500306896

[72] Sivrikaya O, Arol AI. Use of boron compounds as binders in iron ore Pelletization. The Open Mineral Processing Journal. 2014;**3**(1):25-35. DOI: 10.2174/1874841401003010025

[73] Liu H, Xie B, Qin YL. Effect of bentonite on the pelleting properties of iron concentrate. Journal of Chemistry. 2017;**2017**:1-6. DOI: 10.1155/2017/7639326

[74] Liu M et al. Influence of various bentonites on the mechanical properties and impermeability of cement mortars. Construction and Building Materials. 2020;**241**:1-12. DOI: 10.1016/j.conbuildmat.2020.118015

[75] Mohamed OA, Shalabi MEH, El-Hussiny NA, Khedr MH, Mostafa F. The role of normal and activated bentonite on the pelletization of barite iron ore concentrate and the quality of pellets. Powder Technology. 2003;**130**:277-282. DOI: 10.1016/S0032-5910(02)00205-X

Section 2

Environmental and Analytical Applications

Chapter 3

Clay Materials in the Electrochemical Detection of Pollutants

Mariame Coulibaly

Abstract

Clay minerals have been utilized for centuries in diverse domains, including pharmacy, cosmetics, ceramics, and construction. Beyond these traditional uses, clays have attracted considerable interest in the field of chemistry due to their remarkable physicochemical properties, enabling applications as solid catalysts, adsorbents, ion exchangers, and electrochemical sensors. This chapter explores the role of clay materials in electrochemical sensing, with particular emphasis on the mechanisms underlying their function such as surface adsorption, ion exchange, and electron-transfer facilitation. Special attention is given to how these mechanisms contribute to improving the detection of environmental pollutants. Through a detailed examination of recent advances, this chapter highlights the potential of natural clays as low-cost materials whose properties and performance can be further enhanced by various modification processes and additives, depending on the target pollutant. Finally, their advantages as efficient and sustainable modifiers for electrode design in pollutant monitoring systems are discussed.

Keywords: clay minerals, electrochemical sensors, pollutant detection, electrode modification, adsorption mechanisms, electrochemical sensing mechanisms

1. Introduction

Clays are among the most abundant, inexpensive, and versatile natural materials on Earth, and are ubiquitous in soils and sediments [1, 2]. Formed over millions of years through the alteration of rocks, they consist primarily of ultrafine particles (≤ 2 μm) that belong to the phyllosilicate (sheet silicate) family [2, 3]. Their layered crystal structure determines key physicochemical properties, including high specific surface area, significant cation-exchange capacity, and the ability to swell and adsorb various elements [4].

Historically used in pottery production and building materials [5, 6], clays are now employed in diverse industries, including ceramics, paper production, cosmetics, animal feed additives, water treatment, and food contact materials [6, 7]. They are also used as mineral fillers in gels and suspensions for dermocosmetic and biomedical products [8–10].

Beyond these traditional uses, clays have attracted increasing attention in catalysis and environmental remediation owing to their high surface area, ion-exchange

IntechOpen

capacity, and adsorption properties [11–15]. Their structural and surface characteristics are increasingly exploited in electrochemical applications, notably for sensors in heavy metals and organic pollutants detection, electrochemical biosensors, and electrode materials for energy storage and conversion [16–20].

In electrochemical pollutant detection, the interest in clays arises not only from their availability, low cost, and non-toxicity, but also from their ability to immobilize electroactive species, preconcentrate analytes, and in some cases facilitate electron-transfer processes [21]. However, clays also present limitations, including relatively low electrical conductivity and the variability in their composition depending on geological origin [22]. These factors must be considered when optimizing performance for specific uses.

This chapter reviews the use of clay materials in the electrochemical detection of pollutants. First, the structural and physicochemical characteristics relevant to electrode modification are outlined. Next, strategies for preparing clay-based electrodes and approaches for tailoring clays (chemical/structural modification) to enhance sensing performance are discussed. Finally, selected applications in the detection of various contaminants are presented, along with current limitations and future perspectives in this evolving field.

2. Structural and physicochemical properties of clays

Clay science is multidisciplinary, taking in geology, mineralogy, chemistry, materials science, and environmental engineering. This diversity helps explain the lack of consensus on terminology. As noted by Rautureau et al. [23], clay may indicate a fine-grained natural material (generally composed of particles smaller than 2 μm), a soil granulometric fraction, or a plastic material that hardens upon drying or firing. In contrast, clay minerals are hydrated phyllosilicates with layered structures built from SiO_4 tetrahedral sheets and (Al, Mg, Fe) $(O, OH)_6$ octahedral sheets, arranged in 1:1, 2:1, or 2:1:1 type. These minerals are the principal crystalline components of clay materials, although natural clays typically contain a complex mixture of clay minerals and non-clay phases such as quartz, feldspars, or iron oxides [23].

This terminological ambiguity is particularly relevant in electrochemical applications, because both mineralogical structure and physicochemical behavior (surface charge, cation-exchange capacity, swelling, and conductivity) govern performance. Distinguishing the raw material (clay) from its active constituents (clay minerals) is therefore essential when designing and interpreting experiments involving clay-based modified electrodes (CMEs).

2.1 Mineralogical structure of clays

Clay minerals are layered phyllosilicates in which sheets of SiO_4 tetrahedra are linked to sheets of (Al, Mg, Fe) $(O, OH)_6$ octahedra (**Figure 1**) [4, 23]. Three-layer configurations are commonly recognized:

1:1 structure (e.g., kaolinite), where one tetrahedral sheet is bound to one octahedral sheet. Strong hydrogen bonding between layers results in low swelling and limited interlayer ion exchange [4]. Electrochemical activity arises mainly from adsorption on external surfaces, with minimal contribution from internal sites [18, 21].

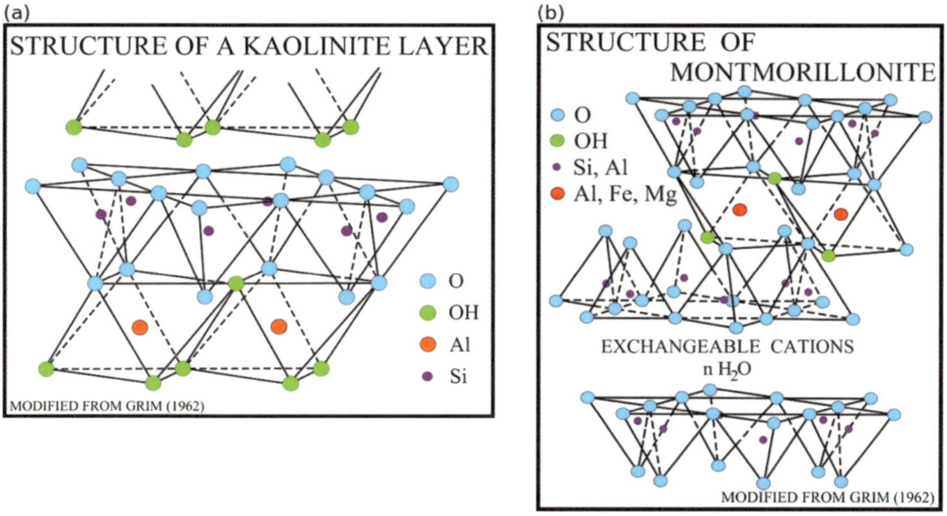

Figure 1.
Layered structures of clay minerals. (a) Kaolinite (1:1, T–O); (b) Montmorillonite (smectite, 2:1, T–O–T). Redrawn from USGS schematics (public domain), modified after Grim (1962) [24].

2:1 structure (e.g., smectites such as montmorillonite, and also illite and vermiculite). Two tetrahedral sheets sandwich one octahedral sheet, and the interlayer region hosts exchangeable cations (Na^+, Ca^{2+}, and K^+) and water [4]. Weaker van der Waals interactions and cation hydration allow significant swelling and expansion, facilitating ion exchange and adsorption [4]. These features make 2:1 clays attractive for electrochemical sensing, as target analytes can be preconcentrated within the interlayers [18, 21].

2:1:1 structure (e.g., chlorite group) [25], in which a 2:1 layer is associated with an additional brucite or gibbsite-like octahedral sheet in the interlayer. This extra sheet strengthens interlayer bonding, producing non-swelling minerals with low cation mobility [26]. Although less favorable for interlayer adsorption, these minerals provide chemically stable external surfaces suitable for immobilizing electroactive species [18, 21].

In summary, the properties of clay minerals depend strongly on their layered structures. In 1:1 clays, hydrogen bonding limits interlayer accessibility, swelling, and cation-exchange capacity (CEC), whereas in 2:1 clays, electrostatic interactions between negatively charged layers and interlayer cations provide reversible ion exchange, with larger interlayer spacing that accommodates water and exchangeable cations. This structural distinction explains differences in reactivity and functional performance among clay minerals [4, 27].

Isomorphic substitution modulates behavior: replacement of framework cations by ions of similar size but different valence (e.g., Al^{3+} for Si^{4+} in tetrahedra; Mg^{2+}/Fe^{2+} for Al^{3+} in octahedra) creates a permanent negative layer charge, balanced by interlayer cations, and directly impacts CEC and adsorption affinity [25, 27]. The location of structural charge (tetrahedral vs. octahedral) also influences electron accessibility and catalytic behavior in modified electrodes [28].

Thus, layered architecture, interlayer chemistry, and charge distribution govern adsorption, ion exchange, interaction with electroactive species, and ultimately their electrochemical performance [4, 27]. For example, 2:1 clays with high CEC and swelling ability enable efficient preconcentration of charged pollutants (e.g., heavy metals, pesticides), enhancing sensor sensitivity [18, 21]. Furthermore, iron-rich clays (e.g., nontronite) can participate in electron transfer, acting as redox mediators. These features help explain why smectites are frequently selected for clay-modified electrodes in pollutant detection [28].

2.2 Physicochemical properties of clay minerals

The properties most relevant for electroanalytical use are surface charge distribution, cation-exchange capacity (CEC), swelling capacity, and specific surface area. Key electrochemically relevant properties are summarized in **Table 1**.

Surface charge distribution: Surface charge has two components: permanent charge from isomorphic substitution within tetrahedral/octahedral sheets (see Section 2.1) and variable charge from pH-dependent protonation-deprotonation at edge sites [25, 27, 29]. The magnitude and location of structural charge control electrostatic attraction/repulsion and influence electron-transfer pathways, thereby affecting adsorption selectivity and signal development [18, 28].

Cation-exchange capacity (CEC): CEC (meq 100 g^{-1}) quantifies the reversible exchange of compensating cations and reflects the density and distribution of permanent charge [25, 27, 30]. High CEC promotes interfacial preconcentration of cationic pollutants (e.g., heavy metals), improving sensitivity; smectites are particularly effective owing to accessible interlayers [30, 31].

Swelling capacity: Swelling is the increase in interlayer spacing driven by hydration of exchangeable cations, which exposes additional internal surfaces and diffusion pathways [32, 33]. In expandable 2:1 clays, this enhances access to adsorption sites and favors intercalation; if excessive, it can compromise film integrity and signal reproducibility [18, 21].

Surface area: Specific surface area (m^2 g^{-1}) comprises external particle surfaces and internal interlayer surfaces; values depend on mineral type, hydration state, and the measurement method (e.g., N_2-BET, EGME) [4, 33]. Large accessible area increases the density of active sites for adsorption and ion exchange, reinforcing the effectiveness of clay-modified electrodes [4, 33].

In summary, electrochemical performance results from the balance between high ion-exchange capacity and large surface area, with appropriate swelling, and is moderated by charge location and film stability.

Property	Typical in clays	Electrochemical implication
Surface charge (permanent + edge)	Permanent (isomorphic substitution) + pH-dependent edge sites	Electrostatic attraction/repulsion
CEC	Low (1:1) High (2:1)	Preconcentration of cations; higher CEC, higher sensitivity
Swelling	Low (1:1) High (2:1)	Interlayer access; excessive swelling may reduce film integrity
Specific surface area	Higher in swelling clays (external + interlayer surfaces)	More active sites; enhanced adsorption/ion exchange

Table 1.
Physicochemical properties of clays relevant to electrochemical sensing and their implications.

3. Preparation and modification of clay-based electrodes

3.1 Pretreatment and purification of clays

Raw clays commonly contain quartz, carbonates, organic matter, and iron oxides, which are typically removed before electrode modification [34, 35]. Pretreatment begins with air-drying, crushing, grinding, and sieving to retain particles below a specific size. When finer fractions are required, the clay is suspended in distilled water at controlled pH, stirred, and fractionated by gravity sedimentation according to Stokes' law [35].

Carbonates are then eliminated using a weak-acid or acetate buffer adjusted near pH ~5 [36, 37]. Homoionization is performed by repeated contact with a salt solution (e.g., Na^+), each step followed by centrifugation; excess electrolyte is removed by dialysis or ultrafiltration until the supernatant conductivity reaches a plateau [34, 38]. When necessary, free iron oxides are selectively dissolved by the dithionite-citrate bicarbonate (DCB) method, which preserves structural iron in the phyllosilicate lattice [39].

Finally, the purified clay is dried and stored in airtight containers. This process improves mineralogical purity, yields a uniform exchange form, and enhances swelling behavior and reproducibility in electrode fabrication [34].

3.2 Modification of clay-based electrodes

Electrode performance depends on clay purity and physicochemical properties, and on the modification strategy. Two main approaches are used: physical deposition of clay films on a conductor and functional modification that integrates or chemically tailors clay within or onto the electrode. The optimal choice is application-dependent (film uniformity, stability, mass transport, and target analyte). **Table 2** summarizes representative modification strategies for clay-based electrodes, listing the substrate, clay type, deposition technique, and any co-modifiers.

3.2.1 Physical deposition methods

Clay dispersed as a colloid or fine powder is applied directly to a conductive substrate to form an adherent, electroactive layer while preserving the natural structure. This approach is simple, requires minimal chemistry, and is compatible with substrates such as glassy carbon, platinum, or indium tin oxide [51–53].

Drop-casting deposits a defined volume of suspension followed by controlled drying; for example, Na-montmorillonite (5 g L^{-1}) films were dried at 100°C for 10 min before use [40].

Spin-coating improves thickness uniformity by high-speed spreading; for example, Na-montmorillonite (12 g L^{-1}) with poly(vinyl alcohol) and colloidal platinum at 8000 rpm for 15 s yielded about 3 μm films [41].

Other options include dip coating, layer-by-layer assembly, spray coating, and electrophoretic deposition, which offer tunable thickness and coverage. Across these techniques, adhesion, porosity, and mechanical stability are governed by suspension composition (solvent, pH/ionic strength), deposition parameters (volume, speed, time, coats), and drying/curing conditions [42, 43, 54].

Electrode (substrate)	Clay type and pretreatment	Deposition technique	Additives/ co-modifiers	Reference
GCE	Na-montmorillonite (Na-exchanged)	Cast film (drop-casting/ evaporation)	—	[40]
GCE	Na-montmorillonite	Spin-coating	PVA (binder), colloidal Pt NPs	[41]
Pt	Two natural clays	Potentiostatic electrodeposition spin-coating (comparative)	—	[42]
GCE	Clay films (natural/ synthetic)	Electrophoretic deposition (EPD)	—	[43]
ITO	Na-montmorillonite	Layer-by-Layer (LbL)	Polyaniline, Au nanoparticles	[44]
GCE	Organoclay (Nafion/clay hybrid)	Cast film	Nafion (perfluoro-sulfonated ionomer)	[45]
GCE	Natural clay	Drop-casting	Chitosan	[46]
Sonogel-carbon (bulk electrode)	Montmorillonite	Composite electrode (sonogel-carbon matrix)	Silica sonogel/ carbon matrix	[47]
GCE	Surfactant-intercalated clays (organoclays)	Cast/assembled films	Quaternary-ammonium intercalants	[48, 49]
GCE/Pt	Laponite intercalated with organosilsesquioxane	Cast film from organosilsesquioxane-laponite sol	Organosilses-quioxane	[50]
CPE (electrode body)	Clay-CPE (clay + graphite + binder)	Mixing/packing	—	[16, 17]

Table 2.
Preparation of clay-based electrodes.

3.2.2 Functional modification methods

Clay is incorporated into the electrode material or chemically bonded to the surface to enhance ion-exchange capacity, adsorption efficiency, and catalytic behavior, while improving robustness.

A widely used approach is the fabrication of clay-carbon paste electrodes (CPEs), where purified clay powders are mixed with graphite powder and a suitable binder to form a homogeneous paste that is packed and polished, providing uniform clay distribution and abundant active sites within a conductive matrix [16, 17].

Beyond CPEs, polymer-clay composites embed clays within conductive or ion-selective polymers (e.g., polyaniline, Nafion, and chitosan) to couple permselectivity and ion exchange with improved charge transport. Layer-by-layer films combining polyaniline, sodium montmorillonite, and Au nanoparticles have enabled simultaneous square-wave anodic stripping voltammetry detection of Pb(II), Cd(II), and Cu(II) at very low levels; Nafion/clay hybrids have been used for paraquat, and chitosan-clay films on glassy carbon have shown high-sensitivity for heavy-metal sensing [44–46].

Sol-gel/sonogel matrices disperse clay platelets within rigid yet porous silica networks, yielding uniform, adherent films with accessible porosity. Clay/sol-gel modified electrodes have been applied to 2,4-dichlorophenol, and sonogel-carbon/montmorillonite platforms have been reported as low-cost, stable supports [47, 55].

Covalent grafting and organoclays employ silane coupling or incorporation of silsesquioxane species to immobilize organic functionalities on or within clay platelets, improving film integrity and tuning permselectivity for ion-exchange-based voltammetry [48, 50, 56]. These chemistries enable stable attachment of recognition elements (e.g., enzymes and chelators) on GCE, Pt, or ITO and can improve selectivity toward target ions or molecules [48, 50, 56].

4. Application in the detection of pollutants

4.1 Inorganic pollutants

Clay-modified electrodes (CMEs) exploit the layered structure and ion-exchange capacity of clays as electrochemical preconcentrators for inorganic pollutants [18, 21, 51]. Detection typically proceeds in two stages: (i) preconcentration (accumulation) at open circuit under stirring, by ion exchange or adsorption (for metals, this step may instead use a controlled deposition potential to reduce the ion), after which the electrode is rinsed, placed in supporting electrolyte, and allowed to rest briefly to eliminate convection [18, 21]; (ii) stripping voltammetry (differential pulse or square-wave) over a potential window that transforms the accumulated species, yielding a peak current proportional to the preconcentrated amount [57, 58]. Sensitivity and selectivity depend on analyte speciation (pH/medium), accumulation time, and film architecture, including interlayer/edge processes or chelation in engineered systems [18, 49, 59].

4.1.1 Speciation and preconcentration

Inorganic pollutants include not only metals but also metalloids, inorganic anions (e.g., nitrate, chromate, and fluoride), and metal/ligand complexes. The speciation of the analyte, whether it exists as a free cation or as an anionic/oxo-complex, together with the working pH, dictates which clay sites govern preconcentration [18, 27]. In 2:1 aluminosilicate clay (e.g., smectite and vermiculite), permanently charged interlayers promote exchange of M^{2+}/M^{3+} and thus strong enrichment of heavy-metal cations at the solid-liquid interface [4]. Edge hydroxyl groups, with pH-dependent charge (often positive at low pH), further contribute through electrostatic adsorption and specific surface complexation [27].

Dominant pathways include: (i) ion exchange and electrostatic adsorption at interlayers and edges, enabling trace-level enrichment of cations and, under certain conditions, selected anions [4, 18, 27]; (ii) ligand-assisted binding, where intercalated ligands provide coordination sites for target ions such as Pb(II), enabling ultra-trace stripping analysis [49, 59]; and (iii) organoclays/grafted ligands, where quaternary-ammonium exchange or covalently attached chelators shifts adsorption equilibria and can improve intrafilm transport and selectivity toward targeted ions or complexes [59].

4.1.2 Electrochemical measurement

After the accumulation step, which can involve either the electrodeposition of a metal or the adsorptive accumulation of a metal-ligand complex (AdSV) without reduction, the analytical signal is recorded during the stripping scan [57, 58]. In anodic or cathodic stripping voltammetry (ASV/CSV), the preconcentrated metal is re-oxidized or re-reduced as potential is swept; in adsorptive stripping voltammetry (AdSV), the adsorbed complex is reduced or oxidized [57, 58]. In all cases, peak current scales with the accumulated amount and thus with bulk concentration, accumulation time, and mass transport within the clay film [18, 57, 58].

For enhanced sensitivity and selectivity, square-wave or differential-pulse voltammetry is commonly used after a brief quiet time in an unstirred supporting electrolyte [57]. For complete optimization, the potential window (vs. reference), pulse parameters (frequency, amplitude, step, and pulse height and width), equilibration time, and any surface regeneration or cleaning potential used between runs [57] should be examined.

4.2 Organic pollutants

Voltammetric detection of organic pollutants at clay-modified electrodes follows the same two-step principle as for inorganic species: an accumulation stage followed

Analyte	Electrode configuration	Technique	LOD ($\mu g\ L^{-1}$)	Real sample	Reference
Pb(II)	ITO/LbL PANI-ES/Au-NPs/Na-MMT	SWASV	0.009	—	[44]
Cd(II)	ITO/LbL PANI-ES/Au-NPs/Na-MMT	SWASV	0.010	—	[44]
Cu(II)	ITO/LbL PANI-ES/Au-NPs/Na-MMT	SWASV	0.020	—	[44]
Zn(II)	Natural clay/chitosan/GCE	SWASV	2.84	Water	[46]
Pb(II)	Natural clay/chitosan/GCE	SWASV	0.891	Water	[46]
Cd(II)	Natural clay/chitosan/GCE	SWASV	2.15	Water	[46]
Cu(II)	Natural clay/chitosan/GCE	SWASV	3.64	Water	[46]
Paraquat	Nafion/clay/GCE	SWCSV	0.5	Real water	[45]
2,4-Dichlorophenol	Clay/sol-gel/GCE	DPV	4.34	—	[55]
2,4-Dichlorophenol	Clay/sol-gel/PGE	DPV	3.15	—	[55]
Zn(II)	Clay-CPE (raw clay)	SWASV	253.02	Drinking water	[16]
Cd(II)	Clay-CPE (raw clay)	SWASV	59.57	Drinking water	[16]
Cu(II)	Clay-CPE (raw clay)	SWASV	30.49	Drinking water	[16]
Acetamiprid	Clay-modified CPE	DPV	21.6	Foodstuffs	[17]
Acetamiprid	Clay-modified CPE	DPV	19.6	Foodstuffs	[17]
Acetamiprid	Clay-modified CPE	DPV	12.0	Foodstuffs	[17]

Table 3.
Analytical performance of clay-modified electrodes.

by a measurement stage. During preconcentration under stirring at open circuit, analyte speciation, defined by charge governed by pK and protonation state as well as by any complexation, determines which clay domains are active for retention (interlayers, edge sites, or engineered organophilic regions) [18, 48, 49]. Enrichment proceeds by ion exchange, electrostatic adsorption, or hydrophobic interactions. In some cases, a mild *in situ* complexation step is introduced to enhance selectivity [18, 48, 49].

In the measurement stage, the accumulated molecules are transformed during the stripping scan, typically recorded by square-wave or differential-pulse voltammetry. Depending on the analyte and detection mode, the process involves reduction, oxidation, or desorption in adsorptive or cathodic stripping voltammetry, and the peak current is proportional to the amount accumulated [57]. As with inorganic pollutants, sensitivity and selectivity depend on analyte speciation (pH and medium), accumulation time, and on the structural and chemical features of the clay film. Functional modifications that increase organophilicity or introduce specific ligands can strengthen affinity for target compounds and improve preconcentration efficiency, with recent examples reported for pesticides such as acetamiprid [17, 45, 48, 49].

Table 3 compiles representative analytical performance data for some clay-modified electrodes.

5. Conclusion and outlook

Clay-modified electrodes combine low cost, robustness, and intrinsic preconcentration capacity. Layered structures operate as selective ion exchange and adsorption media that amplify voltammetric signals without complex instrumentation [18, 21]. Their chemistry can be tailored by pillaring, delamination, organophilization, or intercalation of ligands and macrocycles, which allows a single platform to be adapted for metal cations, oxyanions, and organic pollutants [48, 50].

Several limitations remain: intrinsic film resistivity, restricted mass transport within the film, limited anion handling in many aluminosilicates, and practical issues of reproducibility in complex matrices.

Promising directions include conductive and hierarchical hybrids based on layered double hydroxides, graphene, carbon nanotubes, or conductive oxides, coupled with engineered recognition elements such as ionophores, β-cyclodextrin, calixarenes, and molecularly imprinted polymers, which together can increase sensitivity and selectivity [14, 19]. These advances are particularly suitable for disposable, field-deployable screen-printed formats [59].

From a methodological standpoint, clearer reporting of film loading and thickness, porosity-related parameters, and systematic interference testing would improve reproducibility and enable cross-study comparisons. Standardized definitions for calibration, limits of detection and quantification, stability, and storage conditions should be included in future work.

In summary, clays provide chemically tunable preconcentration media. When combined with conductive scaffolds and appropriate control of analyte speciation, clay-based sensors can evolve into reliable and economical tools for real-world monitoring of inorganic and organic pollutants.

Author details

Mariame Coulibaly
Ecole Normale Supérieure, Abidjan, Côte d'Ivoire

*Address all correspondence to: coulibaly.mariame@ensabj.ci

References

[1] Murray HH. Applied Clay Mineralogy: Occurrences, Processing and Applications of Kaolins, Bentonites, Palygorskite-Sepiolite and Common Clays. Amsterdam: Elsevier; 2007

[2] Foley NK. Environmental Characteristics of Clays and Clay Mineral Deposits. Reston (VA): U.S. Geological Survey; 1999. Available from: https://pubs.usgs.gov/info/clays/ [Accessed: August 22, 2025]

[3] Encyclopaedia Britannica. Phyllosilicate (Sheet Silicate). 2018. Available from: https://www.britannica.com/science/phyllosilicate [Accessed: August 21, 2025]

[4] Bergaya F, Lagaly G, editors. Handbook of Clay Science. 2nd ed. Amsterdam: Elsevier; 2013

[5] Wu X, Zhang C, Goldberg P, et al. Early pottery at 20,000 years ago in Xianrendong Cave, China. Science. 2012;**336**(6089):1696-1700. DOI: 10.1126/science.1218643

[6] Murray HH. Traditional and new applications for kaolin, smectite, and palygorskite: A general overview. Applied Clay Science. 2000;**17**(5-6):207-221. DOI: 10.1016/S0169-1317(00)00016-8

[7] Harvey CC, Lagaly G. Industrial applications. In: Bergaya F, Lagaly G, editors. Handbook of Clay Science. 2nd ed. Amsterdam: Elsevier; 2013. pp. 451-490. DOI: 10.1016/B978-0-08-098259-5.00018-4

[8] Viseras C, Aguzzi C, Cerezo P, López-Galindo A. Uses of clay minerals in semisolid health care and therapeutic products. Applied Clay Science. 2007;**36**(1-3):37-50. DOI: 10.1016/j.clay.2006.07.006

[9] Nomicisio C, Ruggeri M, Bianchi E, et al. Natural and synthetic clay minerals in the pharmaceutical and biomedical fields: A review. Pharmaceutics. 2023;**15**(5):1368. DOI: 10.3390/pharmaceutics15051368

[10] Carretero MI, Pozo M. Clay and non-clay minerals in the pharmaceutical industry: Part I. Excipients and medical applications. Applied Clay Science. 2009;**46**(1):73-80. DOI: 10.1016/j.clay.2009.07.017

[11] Ewis D et al. Adsorption of organic water pollutants by clays and clay-minerals composites: A comprehensive review. Applied Clay Science. 2022;**229**:106686. DOI: 10.1016/j.clay.2022.106686

[12] Khan S et al. Clay-based materials for enhanced water treatment: Adsorption mechanisms, challenges and future directions. Discover Water. 2023;**3**:00083. DOI: 10.1007/s43994-023-00083-0

[13] Zhu L, Tian S, Shi Y. Adsorption of volatile organic compounds onto porous clay heterostructures based on spent organobentonites. Clays and Clay Minerals. 2005;**53**(2):123-136. DOI: 10.1346/CCMN.2005.0530202

[14] Vaccari A. Preparation and catalytic properties of cationic and anionic clays. Catalysis Today. 1998;**41**(1-3):53-71. DOI: 10.1016/S0920-5861(98)00038-8

[15] Macías-Quiroga IF, Rengifo-Herrera JA, Arredondo-López SM. Research trends on pillared interlayered clays (PILCs) used as catalysts in environmental and chemical processes: Bibliometric analysis. The Scientific World Journal. 2022;**2022**:5728678. DOI: 10.1155/2022/5728678

[16] Sanou A, Coulibaly M, N'Dri SR, Tămaș TL, Bizo L, Frentiu T, et al. Raw clay material-based modified carbon paste electrodes for sensitive heavy metal detection in drinking water. Journal of Materials Science. 2024;**59**:13961-13977. DOI: 10.1007/s10853-024-09945-2

[17] Soro SB, Coulibaly M, Gauly LP, N'Dri SR, Bamba D, Trokourey A. Electrochemical detection of acetamiprid in foodstuffs using clay-modified electrodes. Chemistry Africa. 2025;**8**:2155-2167. DOI: 10.1007/s42250-025-01314-z

[18] Mousty C. Sensors and biosensors based on clay-modified electrodes—New trends. Applied Clay Science. 2004;**27**(3-4):159-177. DOI: 10.1016/j.clay.2004.06.005

[19] Lan Y, Liu Y, Li J, Chen D, He G, Parkin IP. Natural clay-based materials for energy storage and conversion applications. Advanced Science. 2021;**8**:2004036. DOI: 10.1002/advs.202004036

[20] Hamidouche F, Ghebache Z, Leprêtre J-C, Djelali N-E. Montmorillonite/poly (pyrrole) for low-cost supercapacitor electrode hybrid materials. Polymers. 2024;**16**(7):919. DOI: 10.3390/polym16070919

[21] Fitch A. Clay-modified electrodes: A review. Clays and Clay Minerals. 1990;**38**(4):391-400. DOI: 10.1346/CCMN.1990.0380408

[22] Paiva LB, Morales AR, Valenzuela Díaz FR. Organoclays: Properties, preparation and applications. Applied Clay Science. 2008;**42**(1-2):8-24. DOI: 10.1016/j.clay.2008.02.006

[23] Rautureau M, Figueiredo Gomes CdS, Liewig N, Katouzian-Safadi M. Clay and clay mineral definition. In: Rautureau M, Figueiredo Gomes CdS, Liewig N, Katouzian-Safadi M, editors. Clays and Health: Properties and Therapeutic Uses. Cham: Springer; 2017. pp. 5-31. DOI: 10.1007/978-3-319-42884-0_2

[24] U.S. Geological Survey (USGS). Clay mineral structure schematics (public domain); modified after: Grim RE. Clay Mineralogy. McGraw-Hill; 1962. Available from: https://pubs.usgs.gov/of/2001/of01-041/htmldocs/clays/smc.htm [Accessed: October 12, 2025]

[25] Brigatti MF, Galán E, Theng BKG. Chapter 2: Structures and mineralogy of clay minerals. In: Bergaya F, Theng BKG, Lagaly G, editors. Handbook of Clay Science. 1st ed. Amsterdam: Elsevier; 2006. pp. 19-86

[26] Martin RT. Reference chlorite characterization for cation exchange capacity. Clays and Clay Minerals. 1954;**3**(1):113-119. DOI: 10.1346/CCMN.1954.0030113

[27] Sposito G. Surface geochemistry of the clay minerals. Proceedings of the National Academy of Sciences of the United States of America. 1999;**96**(7):3358-3364. DOI: 10.1073/pnas.96.7.3358

[28] Xiang Y, Villemure G. Electrodes modified with synthetic clay minerals: Evidence of direct electron transfer from structural iron sites in the clay lattice. Journal of Electroanalytical Chemistry. 1995;**381**:21-27. DOI: 10.1016/0022-0728(94)03629-H

[29] Strawn DG. Sorption mechanisms of chemicals in soils. Soil Systems. 2021;**5**(1):13. DOI: 10.3390/soilsystems5010013

[30] University of Georgia Cooperative Extension. Cation Exchange Capacity and Base Saturation. Circular C1040. Athens (GA): UGA; 2022. Available from:

https://extension.uga.edu/publications/detail.html?number=C1040 [Accessed: August 21, 2025]

[31] Zeyen N, Wang B, Wilson S, Paulo C, Stubbs AR, Power IM, et al. Cation exchange in smectites as a new approach to mineral carbonation. Frontiers in Climate. 2022;**4**:913632. DOI: 10.3389/fclim.2022.913632

[32] Laird DA. Influence of layer charge on swelling of smectites. Applied Clay Science. 2006;**34**(1-4):74-87. DOI: 10.1016/j.clay.2006.01.009

[33] Cases JM, Bérend I, François M, Uriot JP, Michot LJ, Thomas F. Mechanism of adsorption and desorption of water vapor by homoionic montmorillonite: 3. The Mg^{2+}, Ca^{2+}, Sr^{2+} and Ba^{2+} exchanged forms. Clays and Clay Minerals. 1997;**45**(1):8-22. DOI: 10.1346/CCMN.1997.0450102

[34] Moore DM, Reynolds RC Jr. X-Ray Diffraction and the Identification and Analysis of Clay Minerals. 2nd ed. Oxford: Oxford University Press; 1997. 378 p

[35] Gee GW, Bauder JW. Particle-size analysis. In: Klute A, editor. Methods of Soil Analysis. Part 1: Physical and Mineralogical Methods. 2nd ed. Madison (WI): ASA & SSSA; 1986. pp. 383-411

[36] Jackson ML. Soil Chemical Analysis. Englewood Cliffs (NJ): Prentice-Hall; 1958

[37] Schultz MK, Biegalski SR, Inn KGW, et al. Optimizing the removal of carbon phases in soils and sediments for sequential chemical extractions by coulometry. Journal of Environmental Monitoring. 1999;**1**:183-190. DOI: 10.1039/A900534J

[38] Schramm LL. Application of ultrafiltration/dialysis to the preparation of clay suspensions. Clays and Clay Minerals. 1980;**28**(1):67-69. DOI: 10.1346/CCMN.1980.0280111

[39] Mehra OP, Jackson ML. Iron oxide removal from soils and clays by a dithionite-citrate system buffered with sodium bicarbonate. Clays and Clay Minerals. 1960;7:317-327. DOI: 10.1346/CCMN.1958.0070122

[40] Fitch A, Lavy-Feder A, Lee SA, Kirsh MT. Montmorillonite face surface associated tris(bipyridine)chromium (3+) monitored electrochemically. The Journal of Physical Chemistry. 1988;**92**:6665-6670. DOI: 10.1021/j100334a035

[41] Bard AJ, Mallouk TE. Electrodes modified with clays, zeolites, and related microporous solids. In: Murray RW, editor. Molecular Design of Electrode Surfaces. New York: Wiley; 1992. pp. 271-312

[42] Issa AA, Al-Degs YS, Al-Rabady NA. Deposition of two natural clays on a Pt surface using potentiostatic and spin-coating techniques: A comparative study. Clay Minerals. 2008;**43**(3):501-510. DOI: 10.1180/claymin.2008.043.3.13

[43] Song C, Villemure G. Preparation of clay-modified electrodes by electrophoretic deposition of clay films. Journal of Electroanalytical Chemistry. 1999;**462**(2):143-149. DOI: 10.1016/S0022-0728(98)00397-0

[44] De Barros A, da Cruz NC, Constantino CJL, Ferreira M. High performance of electrochemical sensors based on LbL films of gold nanoparticles, polyaniline and sodium montmorillonite clay mineral for simultaneous detection of metal ions. Electrochimica Acta. 2017;**235**:700-708. DOI: 10.1016/j.electacta.2017.03.135

[45] Zen J-M, Jeng S-H, Chen H-J. Determination of paraquat by square-wave voltammetry at a perfluorosulfonated ionomer/clay-modified electrode. Analytical Chemistry. 1996;**68**(3):498-502. DOI: 10.1021/ac950590x

[46] Fort CI, Turdean GL, et al. Green modified electrode for sensitive simultaneous heavy-metal ions electrodetection: Natural clay immobilised with chitosan on glassy carbon. Sensors and Actuators B: Chemical. 2024;**418**:136326. DOI: 10.1016/j.snb.2024.136326

[47] Achache M, Bouchta D, Draoui K, Choukairi M. Low-cost electrochemical sensor based on montmorillonite for the simultaneous detection of phenolic pollutants in tap and bottled water. Journal of Environmental Chemical Engineering. 2025;**13**(2):115981. DOI: 10.1016/j.jece.2025.115981

[48] Tonle IK, Ngameni E, Walcarius A. From clay- to organoclay-film modified electrodes: Tuning charge selectivity in ion-exchange voltammetry. Electrochimica Acta. 2004;**49**(20):3435-3443. DOI: 10.1016/j.electacta.2004.03.012

[49] Ngameni E, Dedzo GK, Walcarius A. Permselective and preconcentration properties of surfactant-intercalated clays: Effect on the electrochemical detection of anionic species. Electroanalysis. 2006;**18**(22):2243-2250. DOI: 10.1002/elan.200603654

[50] Coche-Guérente L, Desprez V, Labbé P. Characterization of organosilasesquioxane-intercalated laponite-clay modified electrodes and (bio)electrochemical applications. Journal of Electroanalytical Chemistry. 1998;**458**(1-2):73-86. DOI: 10.1016/S0022-0728(98)00342-8

[51] Zen J-M, Kumar AS. Peer reviewed: The prospects of clay mineral electrodes. Analytical Chemistry. 2004;**76**:205A-211A. DOI: 10.1021/ac041566z

[52] Ege D, Ghosh PK, White JR, Equey J-F, Bard AJ. Clay-modified electrodes. 3. Electrochemical and electron spin resonance studies of montmorillonite layers. Journal of the American Chemical Society. 1985;**107**:5644-5652. DOI: 10.1021/ja00306a009

[53] Zhou C-H, Shen Z-F, Liu L-H, Liu S-M. Preparation and functionality of clay-containing films. Journal of Materials Chemistry. 2011;**21**:15132-15153. DOI: 10.1039/C1JM11479D

[54] Hayat A, Marty JL. Disposable screen printed electrochemical sensors: Tools for environmental monitoring. Sensors. 2014;**14**(6):10432-10453. DOI: 10.3390/s140610432

[55] Özsoz M, Erdem A, Özkan D, Kerman K, Pinnavaia TJ. Clay/sol-gel-modified electrodes for the selective electrochemical monitoring of 2,4-dichlorophenol. Langmuir. 2003;**19**(11):4728-4732. DOI: 10.1021/la034116v

[56] Guimarães AMF et al. Surface modification of synthetic clay aimed at biomolecule immobilization: Silylation and related strategies (review). Materials Research. 2007;**10**(1):11-20. DOI: 10.1590/S1516-14392007000100004

[57] Mirčeski T, Komorsky-Lovrić Š, Lovrić M. Square-Wave Voltammetry: Theory and Application. Berlin: Springer; 2007. DOI: 10.1007/ 978-3-540-73740-7

[58] Bard AJ, Faulkner LR. Electrochemical Methods: Fundamentals and Applications. 2nd ed. New York: Wiley; 2000. ISBN: 978-0-471-04372-7. 864 p

[59] Wang J. Analytical Electrochemistry. 3rd ed. Hoboken (NJ): Wiley-VCH; 2006. 224 p. ISBN: 978-0-471-70701-8

Chapter 4

Progress in Organic Reactions Catalyzed by Kaolin Clay

Zhi Hong, Ling Liang, Qiwen Wu, Huicong Chen and Wenjing Wang

Abstract

Kaolin is an important clay mineral widely distributed worldwide with extremely abundant reserves on the Earth. It is a 1:1 type double octahedral layered hydrated aluminosilicate clay with two different interlayer surfaces: one is an alumina trihydrate-like configuration where aluminum atoms are octahedrally coordinated with oxygen atoms and hydroxyl groups, and the other is a silica-like framework in which silicon atoms are tetrahedrally coordinated with oxygen. These alternating layers are interconnected *via* hydrogen bonds between aluminum (Al-OH) and siloxane (Si-O) groups. As a result, silica/oxygen and alumina/hydroxyl sheets are exposed and able to interact with various molecules, enabling Kaolin to be readily functionalized with organic groups through host-guest interactions. It can be used as a catalyst or carrier for reactions such as condensation, oxidation, esterification, alkylation, and catalytic cracking. Kaolin is used as a chemical additive material in industries such as ceramics, concrete, papermaking, rubber, plastics, coatings, and refractory materials. In addition, it can be modified through techniques such as thermal activation, acid activation, and intercalation, and is widely used in fields such as environmental protection, materials, and medicine. For example, activated Kaolin serves as an efficacious adsorbent for the removal of heavy metal ions. Interlayer modified Kaolin can be used to prepare high-performance composite materials; Highly active metakaolin prepared by dehydration can improve the performance of hardened concrete. This chapter will systematically explain its physical and chemical properties, modification design, catalytic mechanism, and application examples, providing more theoretical support for the functionalization and sustainable development of mineral materials.

Keywords: Kaolin, natural clay material, modified or unmodified, catalyst or support, organic reaction

1. Introduction

The term Kaolin comes from Jailing Mountain in the eastern suburbs of Jingdezhen, China [1]. Kaolin is an important industrial clay mainly composed of kaolinite, which belongs to natural layered silicate minerals. It is one of the earliest developed and most widely used non-metallic minerals, distributed around the world, and has extremely abundant reserves on Earth [2, 3]. At present, more than

60 countries in the world have Kaolin, and the world's proven reserves of Kaolin are about 32 billion tons. Uzbekistan, China, India, the United States, the Czech Republic, and Türkiye are all major producers [4]. In 2024, the global Kaolin market is expected to reach a size of 4.21 billion US dollars, and is expected to grow from approximately 4.4 billion US dollars in 2025 to 6.28 billion US dollars in 2032, with a compound annual growth rate of 5.2% during the forecast period [5].

The unique mineralogy, morphology, chemistry, and physical specifications of Kaolin make it a versatile raw material suitable for many different applications, such as ceramics, concrete, papermaking, rubber, plastics, fillers, coatings, refractory materials, cosmetics, crayons, fertilizers, detergents, insecticides, white cement, inks, and other industries, as a chemical additive material [6]. In addition, it can be modified through techniques such as thermal activation, acid activation, and intercalation, and is widely used in fields such as environmental protection, materials, and medicine [7–10]. This chapter will systematically explain its physical and chemical properties, modification design, catalytic mechanism, and application examples, providing more theoretical support for the functional application and sustainable development of Kaolin.

2. Properties and preparation methods

2.1 Physical and chemical properties of Kaolin

Kaolin is a white inorganic mineral with a specific gravity of 2.50~2.67, a refractive index of 1.55~1.57, a Mohs hardness of 2.5, and a melting point of 1700°C, and a pH value generally between 6.0 and 8.0. The chemical formula is $Al_2O_3{\cdot}2SiO_2{\cdot}2H_2O$, with a molecular weight of 258.16. It is composed of 46.54% SiO_2, 39.50% Al_2O_3, and 13.96% H_2O. The water here is crystalline water, a 1:1 type double octahedral layered hydrated aluminosilicate clay, with two different interlayer surfaces: one is trihydrate alumina-like, where aluminum atoms coordinate with oxygen atoms and hydroxyl octahedra, and the other is silica-like, where silicon atoms coordinate with oxygen atoms tetrahedrally [11–13]. The adjacent unit layers are bonded by hydrogen bonds, and the interlayer spacing is approximately 7.2 Å (1 Å = 0.1 nm). Kaolin is a salt formed by a weak acid (silicic acid) and a weak base (aluminum hydroxide), and the ratio of the two affects the pH value of its aqueous suspension, generally between 6.0 and 8.0. There is also a type of Kaolin, which contains four crystal waters and has a chemical formula of $Al_2O_3{\cdot}2SiO_2{\cdot}4H_2O$. It is also known as hydrated Kaolin and is a tubular crystal [14–16].

The morphology and aggregation trend of kaolinite can be observed through scanning electron microscopy (SEM) (**Figure 1**) and transmission electron microscopy (TEM) [9]. Kaolin crystals typically have a hexagonal sheet-like structure, with single crystal thicknesses ranging from tens to hundreds of nanometers and lateral dimensions in the micrometer range [17]. These basic layers are combined in various ways to form rich and diverse secondary structures, such as layered, tubularillite (a mineral of the Jailing Tu ethnic group), and blocky. Numerous studies have shown that by appropriately modifying Kaolin, its pore structure and morphology can be adjusted, thereby optimizing its performance in specific applications [18, 19].

The pore structure and microstructure of Kaolin are key factors determining its physical and chemical properties. The analysis of nitrogen adsorption and desorption isotherms can be used to study the pore structure characteristics of Kaolin, providing key parameters. Micropores (<2 nm) and mesopores (2~50 nm) account

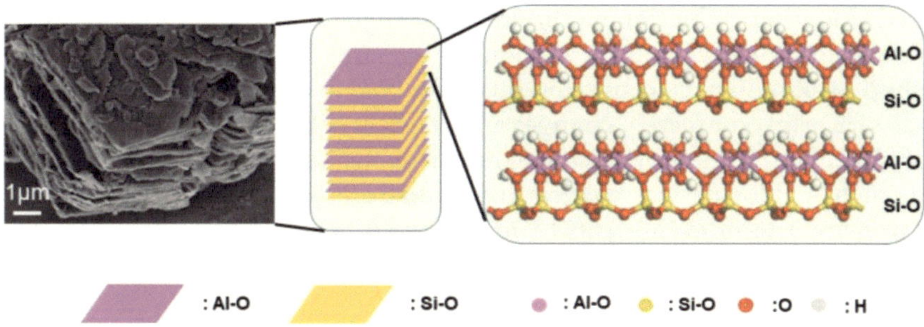

Figure 1.
SEM image, layer structure, and crystal structure of kaolinite.

for the majority of the pore size distribution in Kaolin, with a pore volume of about 0.04~0.18 cm^3/g and a specific surface area of usually 9~27 m^2/g [20, 21]. The loose density of ultrafine Kaolin processed by chemical methods is generally around 0.06~0.07 g/cm^3, while the loose density of ultrafine powder obtained by direct ultrafine grinding of raw ore is about 0.2 g/cm^3 [22].

Chemically speaking, Kaolin is inert and does not undergo chemical reactions with many chemicals. However, due to the presence of adsorbed water, interlayer water, and crystalline water on the surface of Kaolin, there are many active functional groups such as Al-OH bonds, Si-O bonds, etc. In addition, there are many broken chemical bonds at the edges of the crystal lattice, which make the chemical bonds unbalanced and prone to preferential dissolution, dissociation, and adsorption, resulting in some negative charges adhering to the surface of Kaolin [23, 24]. Therefore, the facile functionalization of Kaolin with organic groups is enabled by host-guest interactions. It can be used as a catalyst or carrier for reactions such as condensation, oxidation, esterification, alkylation, and catalytic cracking. In addition, Kaolin has high adhesion, strong acid and alkali resistance, good ion adsorption, and excellent electrical insulation properties. Due to its low cost, large specific surface area, good stability, and potent adsorption capacity, it is extensively employed in environmental treatment and energy conversion.

2.2 Preparation methods

The properties of the source material dictate the processing methods used for Kaolin, the purpose of the product, and the quality requirements of the product. It can be generally divided into three types: purification and whitening, ultrafine processing, and modification.

2.2.1 Purification and whitening

The purification methods of Kaolin include physical methods, chemical methods, and high-temperature calcination methods [5]. The physical purification method mainly utilizes the differences in physical properties, such as particle size, density, magnetism, and surface adsorption of different minerals to separate some non-clay minerals. Common techniques include manual sorting, hydraulic classification, flotation, and magnetic separation. The removal of iron-bearing minerals from Kaolin is commonly achieved through chemical purification and bleaching. Prominent methods consist of acid leaching, chlorination, alkali treatment, reduction, oxidation, and integrated oxidation-reduction methodologies. High-temperature calcination is the best method for removing carbon and whitening, and it is also a necessary processing method for Kaolin in coal-bearing formations. Usually calcined at 650~1050°C, the phase of Kaolin completely changes and releases some impurities, achieving the effect of improving insulation performance. Sometimes, high-temperature treatment methods such as chlorination roasting and oxidation roasting can also be used to remove iron and titanium elements from Kaolin [25, 26].

2.2.2 Ultrafine processing

The main methods for preparing ultrafine Kaolin include mechanical crushing, intercalation, grading, and chemical synthesis [27, 28]. The mechanical crushing method borrows the layered structure characteristics of Kaolin, and under external force, the interlayer force will be destroyed, causing Kaolin to become very small particles, ultimately achieving the goal of ultrafine refining. It is a commonly used method for preparing ultrafine Kaolin. The intercalation method is a technique that involves inserting small molecules into the soil layers of kaolinite using a solution immersion method to break the hydrogen bonds between the layers and expand the interlayer spacing for delamination. It is also known as the intercalation and delamination method and is currently the most promising and effective method for preparing nano-Kaolin. Commonly used small molecules include pyridine, potassium acetate, hydrazine hydrate, urea, dimethyl sulfoxide, etc. The grading method is based on Stokes' law, and the particle size within a certain settling range can be determined by the settling depth of the particles. By settling Kaolin in liquid, ultrafine Kaolin can be obtained. This method has a high cost and low yield, and is not suitable for widespread industrial applications. The chemical synthesis method uses alkaline leaching products of bauxite, sodium metaaluminate, and acidic silica sol products of alkaline acidification to obtain ultrafine synthetic Kaolin through a series of methods. It has excellent properties such as high purity, suspension stability, and light scattering performance, but its synthesis cost is relatively high.

2.2.3 Modification

Natural Kaolin has a low specific surface area, narrow pore channels, and few acidic sites, which are not conducive to the diffusion of substances in the reaction system and the interaction between reactants and acidic sites in Kaolin [2]. Therefore, Kaolin needs to be modified. The main modification methods of Kaolin include acid-base modification, heat treatment, surface modification, intercalation modification, and ion exchange modification [29, 30]. By using these methods to alter the acidity, pore size, specific surface area, surface chemical activity, hydroxyl modification, thermal stability, and polymer compatibility of natural Kaolin, more potential active centers can be provided for catalytic reactions, or it can become a variety of functional materials.

3. Organic reaction catalyzed by kaolin

Natural Kaolin and modified Kaolin can catalyze various organic reactions as acidic or alkaline catalysts. This article reviews various synthesis methods of catalysts supported on Kaolin in recent years, as well as literature on their application in organic synthesis, and evaluates the catalytic activity of these catalysts.

3.1 Organic reaction catalyzed by natural Kaolin

Kaolin could catalyze chemical reactions by acting as Brϕnsted acids or Lewis acids: cyclization, condensation, isomerization, esterification, and ester exchange reaction have been performed with clay catalysts.

3.1.1 Synthesis of aliphatic heterocycles

In 2013, in order to clarify the Biginelli reaction mechanism, a Kaolin-catalyzed solvent-free synthesis of compounds containing 4*H*-pyrimido[2,1-b][1,3]benzothiazole (**1a~1g**) derivatives and 3,4-dihydropyrimidin-2(1*H*)-one/thione (**2a~2b**) was carried out using a one-pot three-component reaction of diketones, aldehydes, and 2-amino benzothiazole, urea, or thiourea. It shows the dependency of the mechanistic behavior on the amine scaffold, which formed the intermediate in the first step of the mechanism (**Figure 2**) [31].

3.1.2 Synthesis of aromatic heterocycles

5-Substituted-1H-tetrazoles are valuable heterocyclic scaffolds that serve as metabolically stable bioisosteres, effectively mimicking carboxyl groups and cis-amide bonds. In 2010, Chermahini et al. discovered that under the condition of 100~130°C, Kaolin can catalyze the reaction of various nitriles with NaN_3, achieving moderate to good yields (**3a~3m**). This catalytic system has the advantages of mild reaction conditions, simple posttreatment, and green efficiency (**Figure 3**) [32].

3.1.3 Synthesis of oxime compounds

In 2019, our research group discovered that superfine Kaolin efficiently catalyzed the synthesis of oxime compounds (**4a~4v**) from carbonyl compounds and hydroxylamine hydrochloride under microwave irradiation, achieving a maximum yield of 92%. Compared to conventional methods, this approach significantly shortens the reaction time, enhances the product yield, and offers simplified work-up procedures along with environmental benignity. Additionally, based on the aforementioned experimental conditions, we conducted a preliminary study on the transoximation reaction and found that superfine Kaolin is also applicable for catalyzing the oxime

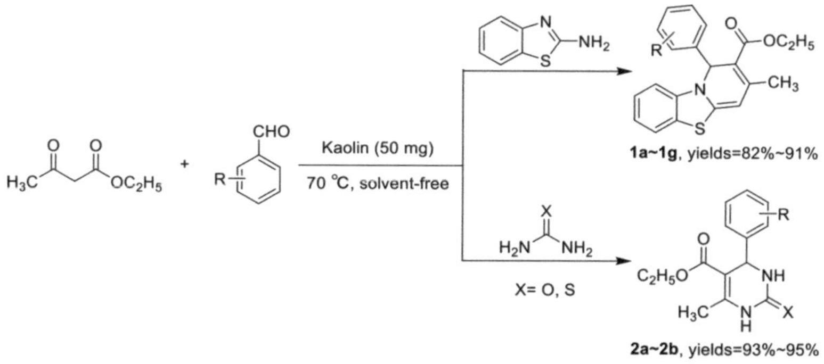

Figure 2.
Natural Kaolin-catalyzed Biginelli-type reaction.

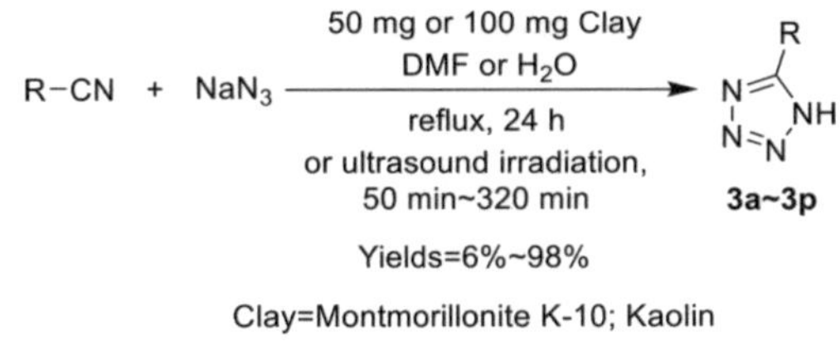

Figure 3.
Natural Kaolin-catalyzed synthesis of 5-substituent 1-H-tetrazoles.

exchange reaction, with yields ranging from 42% to 67% (**4s, 4w~4y**). This further enriches the applications of Kaolin in organic reactions. To this end, we propose the following reaction mechanism: In this condensation reaction, carbonyl group activation is achieved due to the Lewis acidity provided by SiO_2 in superfine Kaolin. This

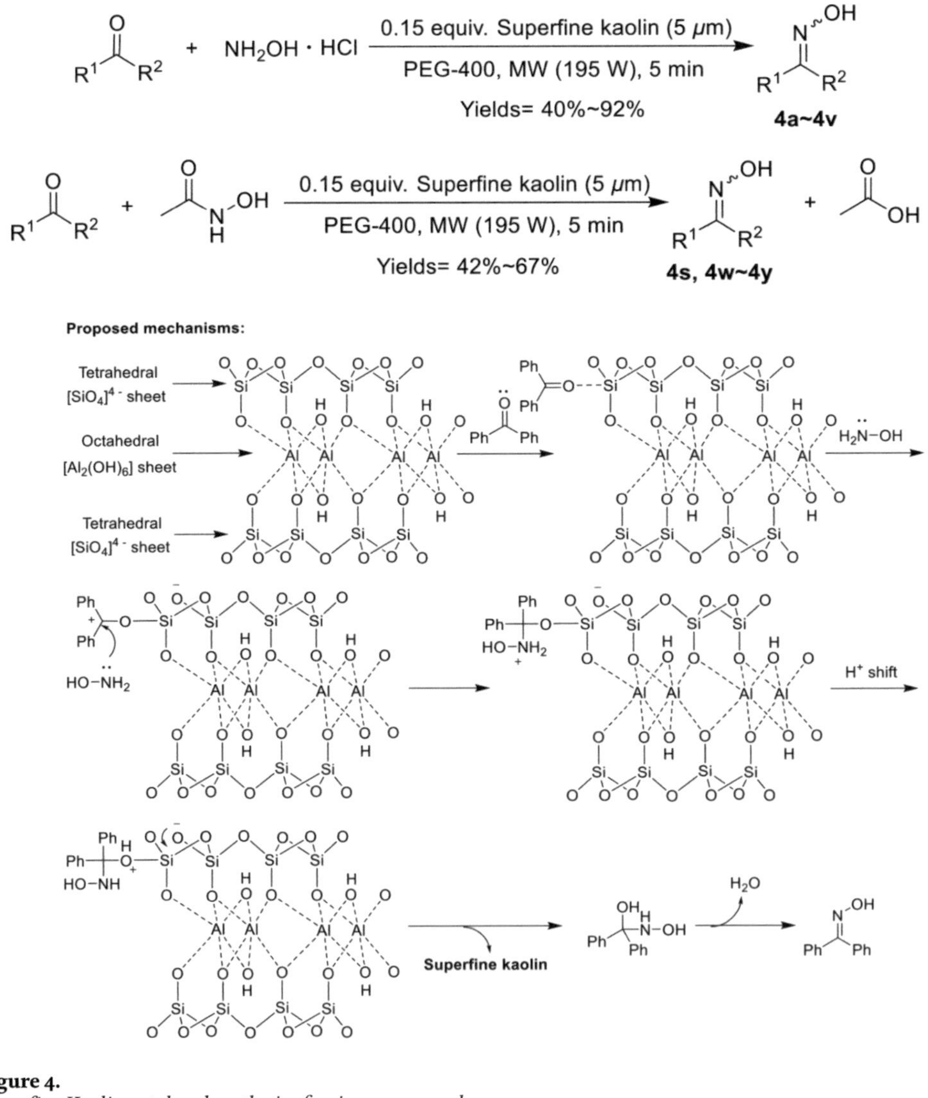

Figure 4.
Superfine Kaolin catalyzed synthesis of oxime compounds.

activation reduces the electron density on the carbonyl carbon atom, thereby facilitating nucleophilic attack by hydroxylamine. The oxime products are subsequently formed through sequential proton transfer and dehydration steps (**Figure 4**) [33].

3.1.4 Kabachnik-fields reaction

In 2024, Boukachabia et al. reported the Kabachnik-Fields reaction catalyzed by superfine Kaolin, in which multi-component condensation reactions were carried out in a one-pot method to synthesize α-aminophosphonic acid ester derivatives (**5a~5n**). To optimize the process, a Box-Behnken design was utilized for establishing reaction conditions, exploring its scope, and successfully scaling the synthesis to the gram level. The results indicate that Kaolin has good catalytic performance, and the presence of electron-withdrawing groups (-NO_2 or -CN) in the meta or para position of aldehydes is beneficial for the reaction (**Figure 5**) [34].

3.1.5 Isomerization reaction

In 2018, Pienkoß's team investigated the glucose-fructose isomerization and its in-situ etherification to methyl fructose (**MF**) in methanol solvent using natural Kaolin as a catalyst. This research demonstrated an effective approach to catalyze the target reaction using this inexpensive (about $0.16 per kilogram) and naturally abundant clay material. Under optimized conditions, a yield of up to 52% of **MF** was achieved. According to the research results, the reaction mechanism involves the bifunctional (Lewis acid and Brønsted acid) catalytic effect of Kaolin, where glucose is isomerized to fructose, and fructose rapidly undergoes etherification at the anomeric carbon position in methanol solvent to form **MF** (**Figure 6**) [35].

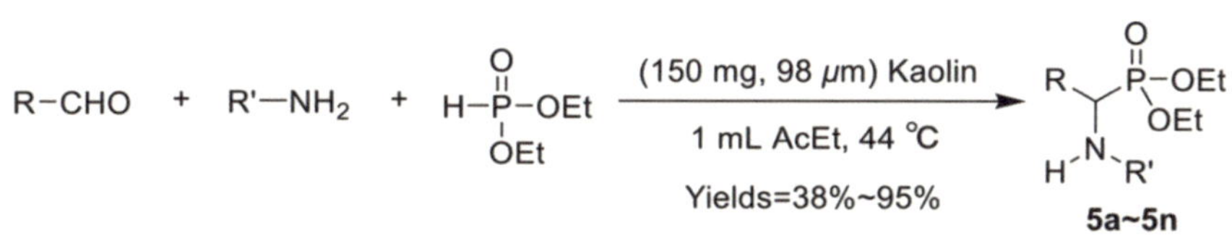

Figure 5.
Superfine Kaolin-catalyzed Kabachnik-Fields reaction.

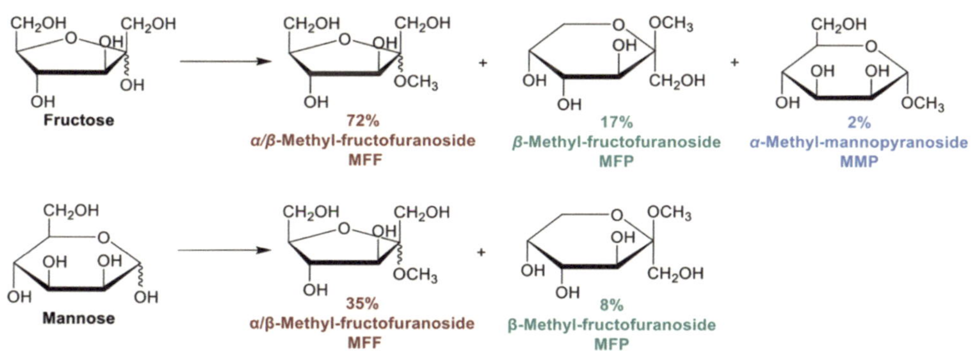

Figure 6.
Reactivity of individual sugars present during glucose isomerization in methanol with Kaolin under continuous-flow operation conditions (120°C, 3 h, 3 wt% sugar in 5 mL methanol, 60 wt% Kaolin relative to sugar, 500 rpm).

3.1.6 Esterification reaction

In 2025, Larba et al. reported the synthesis of novel terpene esters through the esterification of monoterpenes with anhydrides using natural Kaolin as a catalyst, and evaluated their potential against the Ebola virus through computational simulation. This study demonstrated the multifunctionality of natural Kaolin as a low-cost and eco-friendly catalyst, achieving efficient synthesis of various terpene esters (**6a~6i**) (yields = 30% ~ 100%) under optimized conditions (50°C, acetonitrile solvent) (**Figure 7**) [36].

3.2 Organic reaction catalyzed by modified Kaolin

3.2.1 Synthesis of aliphatic heterocycles

In 2019, Mirjalili and Soltani reported the synthesis of pyrimido[2,1-b]benzothiazoles (**7a~7l**) *via* the one-pot condensation of 2-aminobenzothiazole, an aldehyde, and β-keto ester catalyzed by magnetic nano catalyst nano-Kaolin/Ti^{4+}/Fe_3O_4 with middle to excellent yields under solvent-free conditions at 100°C. In this reaction, the Ti^{4+} cation in the catalyst acts as a Lewis acid and activates the carbonyl groups in the substrates to promote the reaction (**Figure 8**). This protocol had some advantages at short reaction times, high yields, easy work-up procedure, and easy separation of the catalyst with reusability [37].

In 2023, Khelili group reported a green method for synthesizing 1,4-dihydropyridines (**8a-8j**) by a one-pot reaction of various aldehydes, ethyl acetoacetate, and ammonium carbonate, catalyzed by a seashell and Kaolin powders, mixed and calcined at 500°C and 800°C, under solvent-free conditions, has been developed (**Figure 9**). The mixture composed of 20% seashell and 80% Kaolin (Q3–500),

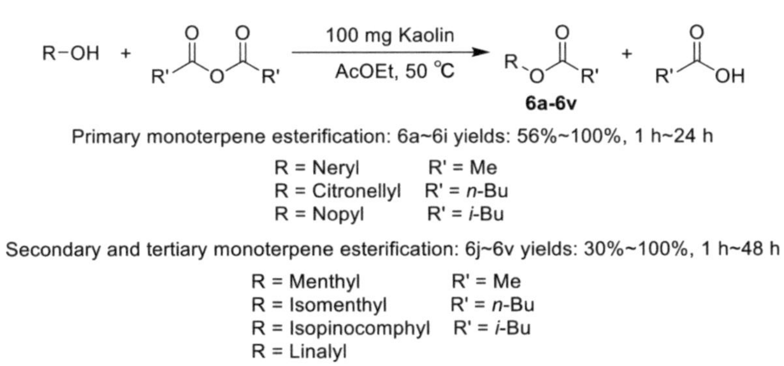

Figure 7.
Natural Kaolin-catalyzed esterification reaction of monoterpene with anhydrides.

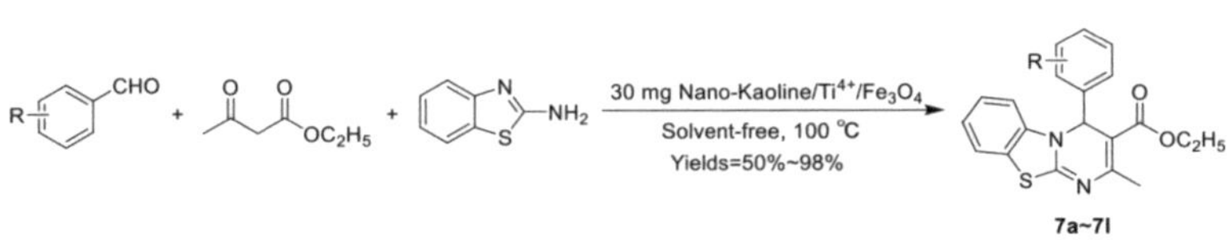

Figure 8.
Synthesis of pyrimido[2,1-b]benzothiazoles catalyzed by nano-Kaolin/Ti^{4+}/Fe_3O_4.

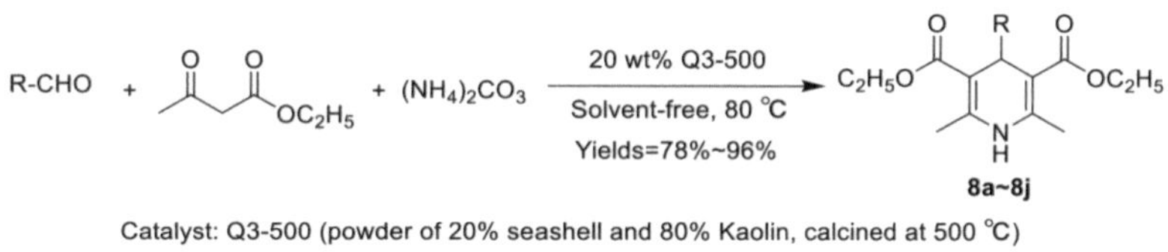

Figure 9.
Synthesis of 1,4-dihydropyridines catalyzed by Seashell and Kaolin powders.

calcined at 500°C, gave the best results (78%~96% yields). Q3–500 separation from the reacting medium was carried out by filtration and reused several times with a slight decrease in yields. Compared with other conventional methods, the present method is inexpensive and offers advantages, namely: the catalyst is eco-friendly, recyclable, and gives high yields with shorter reaction time [38].

3.2.2 Synthesis of aromatic heterocycles

In 2025, Wang et al. developed a novel biocompatible nano catalyst composed of curcumin-mediated and chitosan (CS)-modified Kaolin-supported silver nanoparticles (AgNPs) (Kaolin@CS-Cur/AgNPs). This material can efficiently catalyze the three-component coupling reaction (MCR) for the synthesis of 1-substituted-1H-tetrazole derivatives (**9a~9j**) under solvent-free conditions, with the highest yield reaching 98% (**Figure 10**). Silver nanoparticles catalyze the reaction by coordinating with the oxygen atoms of triethyl orthoformate's ethoxy groups, which activate the carbon center for nucleophilic attack by the amine. Subsequently, the reaction proceeds *via* sequential nucleophilic attack by the azide group and cyclization to afford the desired tetrazole products [39].

3.2.3 Condensation reaction

In 2012, Gordi et al. reported a method for synthesizing bis(indolyl)methanes derivatives (**10a~10m**) by condensation reaction of indole and aldehyde at room temperature using sulfuric acid-modified Kaolin (2% w/w) as an efficient and environmentally friendly catalyst. The catalyst is easy to prepare, cost-effective, and recyclable, with a short reaction time (15~35 minutes) and a yield of up to 80%~95% (**Figure 11**) [40].

3.2.4 Isomerization reaction

In 2024, Sidorenko et al. proposed a reaction mechanism for the isomerization of α-pinene epoxide catalyzed by acid-modified kaolinite nanotubes (KNT-Pir). This mechanism indicates that in non-polar solvents such as cyclohexane, the Lewis

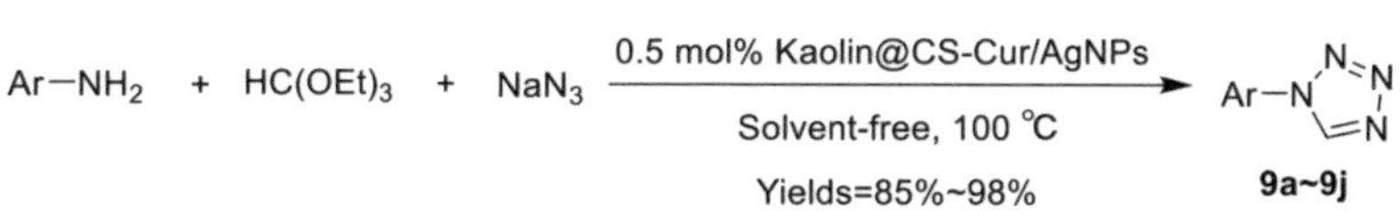

Figure 10.
Synthesis of 1-substituted-1H-tetrazoles catalyzed by Kaolin@CS-Cur/AgNPs.

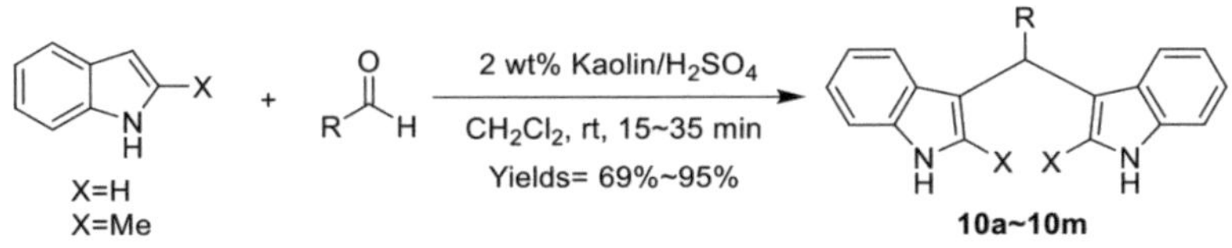

Figure 11.
Synthesis of bis(indolyl)methanes catalyzed Kaolin/sulfuric acid.

acid site (Al^{3+}) dominates the reaction and preferentially catalyzes the protonation of α-pinene epoxide (**11**) to form a carbocation intermediate **(11-A)**. Subsequently, intermediates (**11-B** and **11-C**) were generated by ring opening of cyclobutane, which were ultimately converted into campholenic aldehyde (**12**) and iso-campholenic aldehyde (**13**), with a total selectivity of 57.8%. This process benefits from the directed activation of epoxy bonds by Lewis acids, promoting region-selective cleavage of cyclic structures. In polar solvents such as DMSO, the effect of Brønsted acid sites (-OH) is significantly enhanced. The carbocation intermediate **(11-A)** is rearranged into (**11-D**) under proton transfer, and further converted into trans-carveol (**14**) and trans-sorberol (**15**). Polar solvents stabilize transition states and promote hydrogen transfer, resulting in a high selectivity of 55%~57% for trans-carvacrol (**Figure 12**) [41].

3.2.5 Sonogashira reaction

In 2021, Mahmoud et al. reported the Sonogashira coupling reaction (SCR) between aryl halides and acetylenes under aerobic conditions, catalyzed by a palladium nanoparticle (Pd NPs@CS-Kao) complex supported on a novel Schiff base-modified chitosan-Kaolin composite in ethanol solvent, achieving a maximum product (**16a~16o**) yield of up to 89% (**Figure 13**) [42].

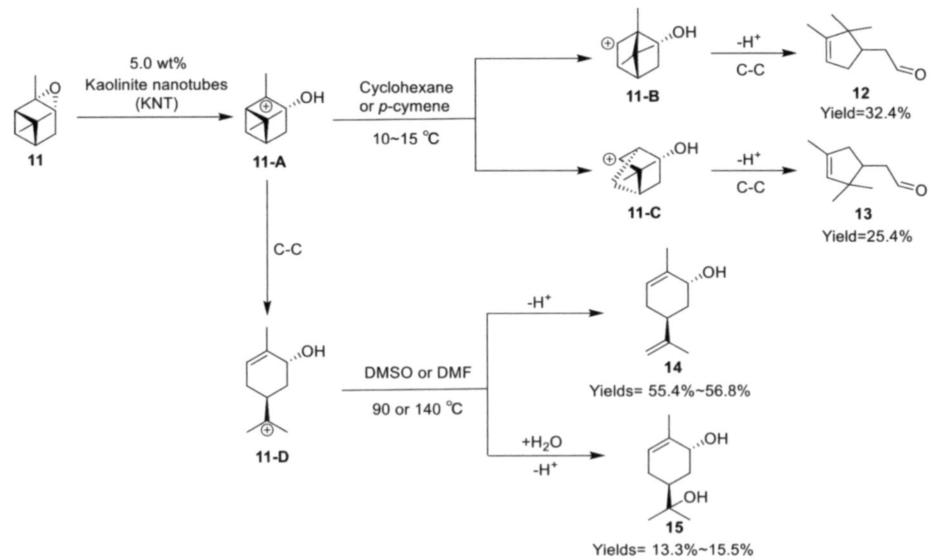

Figure 12.
Isomerization reaction catalyzed by kaolinite nanotubes (KNT).

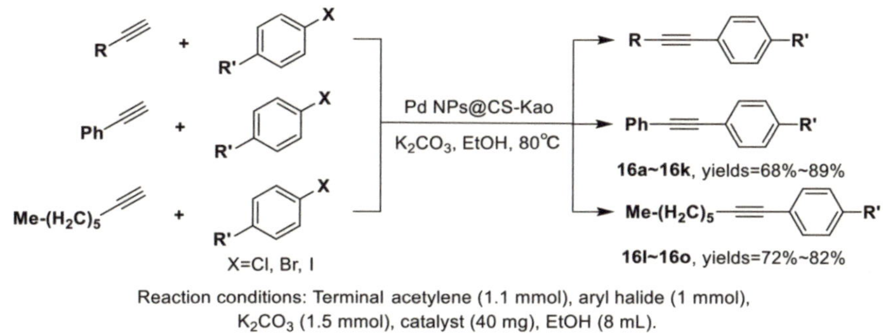

Figure 13.
Pd NPs@CS-Kao catalyzed the SCR of terminal alkynes with different aryl halides.

3.2.6 Suzuki reaction

In 2020, Talat et al. developed a catalyst exhibiting exceptional catalytic activity and reusability by utilizing a Schiff base-modified Kaolin material loaded with Fe_3O_4 as a novel support. This support enabled the immobilization of palladium nanoparticles (Pd NPs@Kao/Fe_3O_4/Pyr) without requiring an additional reducing agent. The catalyst was subsequently employed in the Suzuki cross-coupling reaction for the synthesis of biaryl compounds (**17a**) (**Figure 14**) [43].

A chitosan-pectin composite-functionalized Kaolin (Kaolin@CS-Pectin/Pd) was employed as a novel heterogeneous support for the facile and eco-friendly synthesis of ultrafine palladium nanoparticles *via* a mild and efficient green synthetic protocol. In 2022, Marjan et al. prepared Kaolin@CS-Pectin/Pd and employed it for the efficient synthesis of diverse biaryl compounds (**18a~18m**). Additionally, the catalytic performance of the material was evaluated in the reduction of *p*-nitrophenol. In both protocols, the catalyst exhibited excellent performance (**Figure 15**) [44].

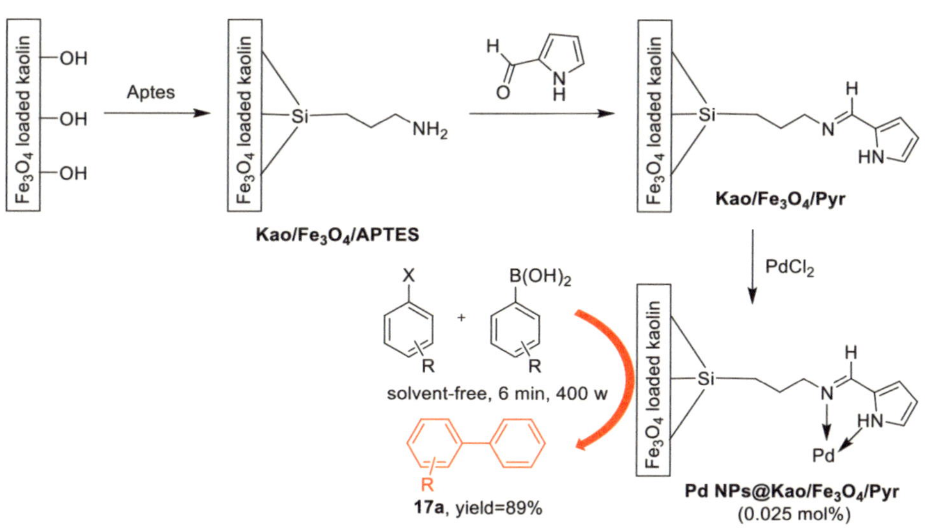

Figure 14.
Synthesis of Pd NPs@Kao/Fe_3O_4/Pyr.

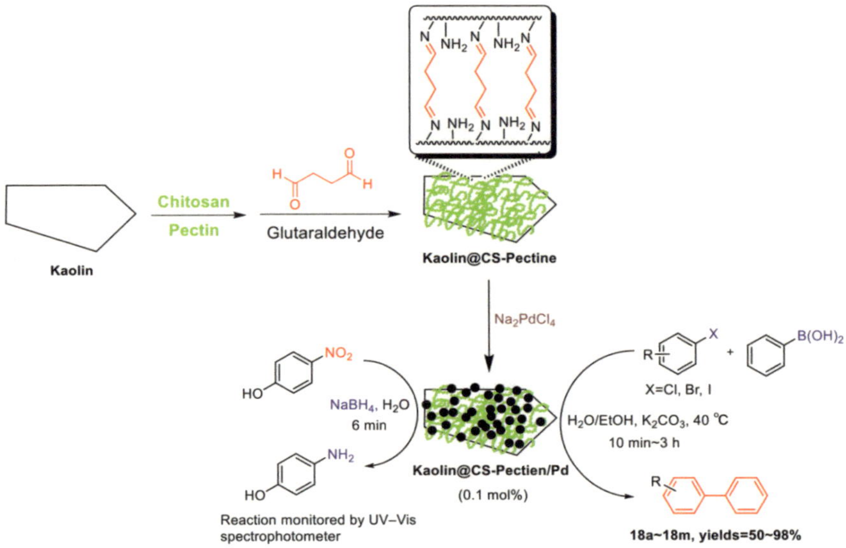

Figure 15.
Synthesis of Kaolin@CS-Pectin/Pd nanocomposite and its catalytic application in the reduction of 4-nitrophenol and Suzuki-Miyaura coupling reactions.

3.2.7 C(sp^3)-H bond activation reaction

In 2021, Bao et al. successfully prepared an iron nanoparticle (FeNP)-loaded kaolinite heterogeneous catalyst using iron salts and kaolinite. The catalytic activity of this catalyst for C(sp^3)-H bond activation in cross-dehydrogenative coupling (CDC) reactions was determined using the direct esterification of benzoic acid with cyclic ether as a model reaction (**19a–19t**, **20a–20f**) (**Figure 16**) [45].

3.2.8 Reduction reaction

In 2023, Gan's group demonstrated that 4-nitrophenol could be efficiently reduced in the presence of bimetallic Pd–Ag alloy nanoparticles supported on polydopamine-functionalized Kaolin (Kaolin-PDA-Pd–Ag), exhibiting a high kinetic rate constant of 3.46 min^{-1} at a concentration of 10 mM under ambient temperature conditions. Furthermore, the Kaolin-PDA-Pd-Ag nanocomposite exhibited excellent catalytic

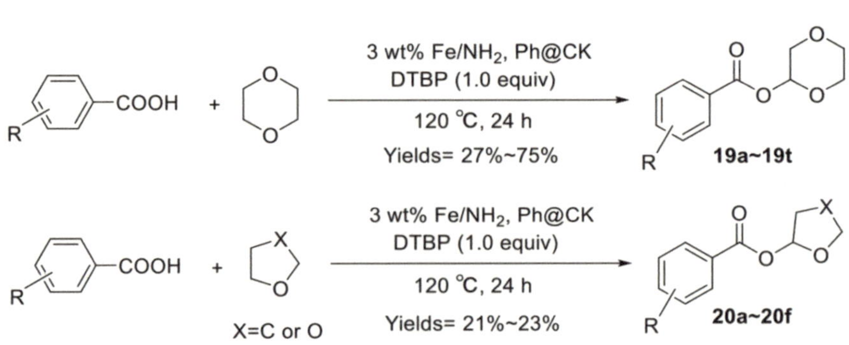

Figure 16.
Substitute the scope of various benzoic acids catalyzed reaction with 1,4-dioxane or 1,3-dioxolane.

performance toward the organic dyes methyl orange and safranin-T. The enhanced catalytic activity of the Kaolin-PDA-Pd-Ag composite can be attributed to the electronic effects and the synergistic interaction inherent in the bimetallic Pd-Ag alloy. These results suggest that Kaolin-PDA-Pd-Ag composites could be applied as high-efficiency catalysts for the reduction of 4-nitrophenol and degradation of organic dyes (**Figure 17**) [46].

Asmare et al. successfully synthesized CuO/Kaolin nanocomposites (CuO/Kaolin NC) *via* a one-pot method. The CuO/Kaolin nanocomposite exhibited excellent catalytic activity in the reduction of 4-nitrophenol (**4-NP**) to 4-aminophenol (**4-AP**) using $NaBH_4$ as a reducing agent under ambient temperature conditions, with a conversion rate exceeding 99% and a rate constant of 1.76 min^{-1}. Additionally, the catalyst exhibited excellent reusability and stability, making it a cost-effective and environmentally benign non-noble metal oxide nano catalyst for the efficient removal of nitro compounds from water (**Figure 18**) [47].

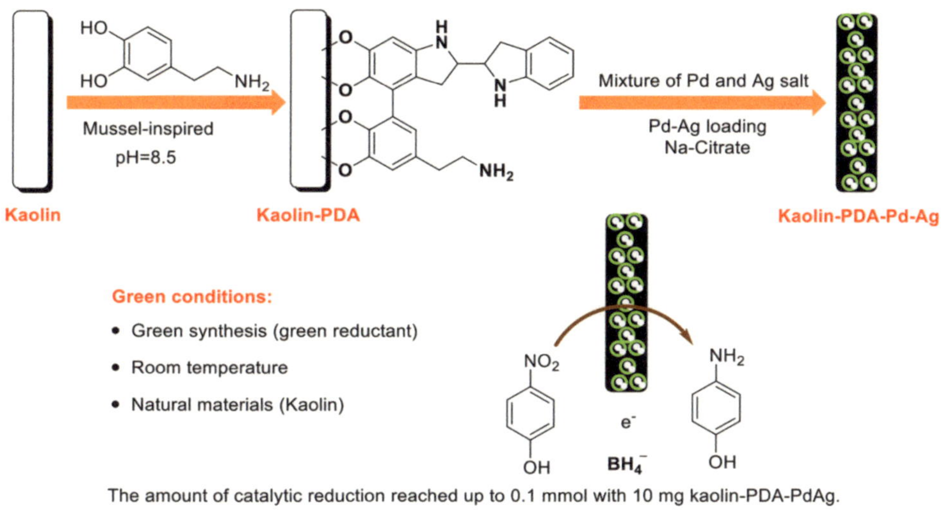

Figure 17.
Schematic illustration of the proposed catalytic mechanism for the reduction of 4-nitrophenol by $NaBH_4$ in the presence of Kaolin-PDA-Pd-Ag composite catalysts.

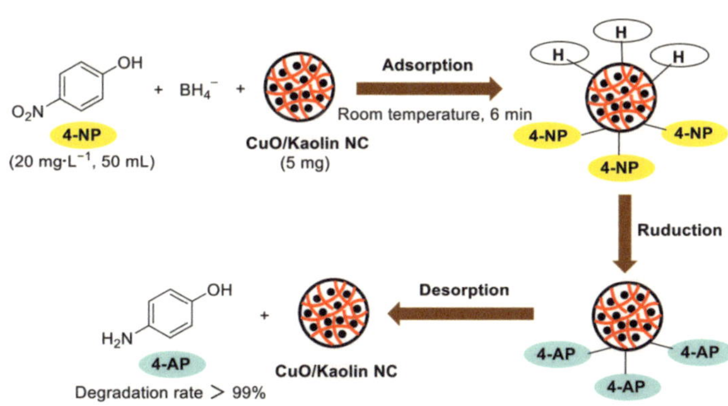

Figure 18.
Schematic representation of the plausible catalytic reduction mechanism of 4-nitrophenol by $NaBH_4$ over the CuO/Kaolin NC.

4. Conclusions

In this review, we emphasize the applications of natural and modified Kaolin in the field of organic synthesis over the past two decades, which can help us better understand the chemistry and behavior of Kaolin. Metals, metal oxides, inorganic complexes, and some organic compounds are constantly being used in the preparation of modified Kaolin catalysts, which provide very attractive possibilities and opportunities for more organic synthesis reactions, producing unique selectivity that other catalysts cannot achieve, as well as a technically attractive zero-waste alternative. Although Kaolin as a catalyst still has some drawbacks, its excellent catalytic activity makes it pale in comparison, which has aroused the interest of many researchers engaged in this research, especially in the synthesis of selective coupling products through C-X bond formation reactions and multi-component reactions.

Acknowledgements

The authors thank the Funding for Basic Public Welfare Research Projects in Zhejiang Province of China (LGG18B020002) for providing financial support for this project.

Conflict of interest

The authors declare no conflict of interest.

Author details

Zhi Hong*, Ling Liang, Qiwen Wu, Huicong Chen and Wenjing Wang
School of Pharmaceutical and Chemical Engineering, Taizhou University, Taizhou, China

*Address all correspondence to: zhihong@tzc.edu.cn

References

[1] Xu J, Wang H. An IR spectroscopy study of kaolinite from Jingdezhen area. Acta Petrologica et Mineralogica. 2010;**29**(1):59-66

[2] Matusik J. Arsenate, orthophosphate, sulfate, and nitrate sorption equilibria and kinetics for halloysite and kaolinites with an induced positive charge. Chemical Engineering Journal. 2014;**246**:244-253

[3] Dill HG. Kaolin: Soil, rock and ore: From the mineral to the magmatic, sedimentary and metamorphic environments. Earth Science Reviews. 2016;**161**:16-129

[4] Yaghmaeiyan N, Mirzaei M, Delghavi R. Introduction of kaolin clay and its application as a heterogeneous catalyst in organic syntheses: A review. Journal of the Iranian Chemical Society. 2025;**22**(4):667-682

[5] Zdretsov IM, Gerasimov AM. Green and low-cost synthesis of zeolites from kaolin: A promising technology or a delusion? Reaction Chemistry & Engineering. 2024;**9**(8):1994-2027

[6] Wang CC, Shi CJ, Huang JR, Wei XF, Shi YF, Xiao LP, et al. Synergistic procoagulant mechanism and application of kaolin-zeolite composite hemostat for effective hemorrhage control. ACS Applied Materials & Interfaces. 2024;**16**(37):49186-49196

[7] Liang X, Li Q, Fang Y. Preparation and characterization of modified kaolin by a mechanochemical method. Materials. 2023;**16**(6):3099

[8] Bakather OY, Zouli N. Enhancing Kaolin's structure for efficient removal of lead ions from aqueous solutions. Materials Today Communications. 2024;**40**:110004

[9] Qu YX, Fan D, Li FJ, Ouyang PW, Fu L, Yang HM. Exfoliating kaolin to ultrathin nanosheets with high aspect ratio and pore volume: A comparative study of three kaolin clays in China. Applied Surface Science. 2023;**635**:157778

[10] Zhang Y, Huang P, Long M, Liu S, Yang H, Yuan S, et al. Intercalated kaolinite as an emerging platform for cancer therapy. Science China. Chemistry. 2019;**62**(1):58-61

[11] Khan MI, Khan HU, Azizli K, Sufian S, Man Z, Siyal AA, et al. The pyrolysis kinetics of the conversion of Malaysian kaolin to metakaolin. Applied Clay Science. 2017;**146**:152-161

[12] Yan K, Guo Y, Fang L, Cui L, Cheng F, Li T. Decomposition and phase transformation mechanism of kaolinite calcined with sodium carbonate. Applied Clay Science. 2017;**147**:90-96

[13] Kogure T. Visualization of clay minerals at the atomic scale. Clay Minerals. 2020;**55**(3):1-16

[14] Ochirkhuyag A, Temuujin J. The catalytic potential of modified clays: A review. Minerals. 2024;**14**(6):629

[15] Li Y, Liu Q, Liu L, Liu L, Hou D, Wu Y. Effect of original crystal size of kaolinite on the formation of intercalation compounds of coal-measure kaolinite. Materials Today Communications. 2023;**35**:106130

[16] Fan J, Liu X, Liu M, Yang M, Jiang Y, Mi R, et al. Characterization of thermal behavior of two types of kaolin in China

by ultrafast joule heating combined with XRD, FT-IR, TG-DSC and SEM. Thermochimica Acta. 2024;**742**:179894

[17] Ngnie G, Dedzo GK, Detellier C. Synthesis and catalytic application of palladium nanoparticles supported on kaolinite-based nanohybrid materials. Dalton Transactions. 2016;**45**:9065-9072

[18] Castellano M, Turturro A, Riani P, Montanari T, Finocchio E, Ramis G, et al. Bulk and surface properties of commercial Kaolins. Applied Clay Science. 2010;**48**:446-454

[19] Bao J, Rao W, Zhou Y, Wen B, Wang B, Lv G, et al. Effect of the microstructure of support materials on cracking catalyst performance. Crystals. 2023;**13**:123

[20] Okada K, Yoshizaki H, Kameshima Y, Nakajima A. Effect of the crystallinity of kaolinite precursors on the properties of mesoporous silicas. Applied Clay Science. 2008;**41**(1-2):10-16

[21] Mustapha S, Tijani JO, Ndamitso MM, Abdulkareem AS, Mohammed AK. Adsorptive removal of pollutants from industrial wastewater using mesoporous kaolin and kaolin/TiO_2 nanoadsorbents. Environmental Nanotechnology, Monitoring and Management. 2021;**15**:100414

[22] Cheng W, Zhu YJ, Shao JA, Zhang WN, Wu GH, Jiang H, et al. Mitigation of ultrafine particulate matter emission from agricultural biomass pellet combustion by the additive of phosphoric acid modified kaolin. Renewable Energy. 2021;**172**:177-187

[23] Wang XB, Xie YA, Wang Q, Shang HL, Hu ZC, Ku JG, et al. Removal and separation of iron and carbon from kaolin: A review. Powder Technology. 2025;**458**:120939

[24] Hartati D, Prasetyoko M, Santoso I, Qoniah WL, Leaw PBD, Firda, et al. A review on synthesis of kaolin-based zeolite and the effect of impurities. Journal of the Chinese Chemical Society. 2020;**67**(6):911-936

[25] Wang CW, Wang YX, Sun W, Liu RQ, Dong YD, Xu R, et al. New insights into the interaction of sodium hexametaphosphate with coal slime and its application in the purification of kaolin. Journal of Cleaner Production. 2024;**446**:140956

[26] Matłok M, Petrus R, Warchoł JK. Equilibrium study of heavy metals adsorption on kaolin. Industrial and Engineering Chemistry Research. 2015;**54**(27):6975-6984

[27] Mu Y, Yang SJ, Shao ZQ, Fu ZK, Ren SL. Preparation of modified kaolin nanoparticle demulsifier and its demulsification performance for crude oil-in-water emulsion. Energy Fuel. 2024;**38**(16):15172-15181

[28] Wei WC, Zhang YY, Fu XH, Yuan SZ, Shi CX, Su ZC, et al. Nano-kaolin enhances the mechanical, electrical, and thermal properties of cellulose insulating paper. High Voltage. 2025;**10**(3):646-659

[29] Zhang Q, Yan Z, Ouyang J, Zhang Y, Yang H, Chen D. Chemically modified kaolinite nanolayers for the removal of organic pollutants. Applied Clay Science. 2018;**157**:283-290

[30] Cristóbal AGS, Castelló R, Luengo MAM, Vizcayno C. Acid activation of mechanically and thermally modified Kaolins. Materials Research Bulletin. 2009;**44**(11):2103-2111

[31] Sahu PK, Sahu PK, Agarwal DD. Efficient and facile synthesis of heterocycles and their mechanistic

consideration using kaolin. RSC Advances. 2013;**25**(3):9854-9864

[32] Chermahini AN, Teimouri A, Momenbeik F, Zarei A, Dalirnasab Z, Ghaedi A, et al. Clay-catalyzed synthesis of 5-substituent 1H-tetrazoles. Journal of Heterocyclic Chemistry. 2010;**47**(4):913-922

[33] Liang L, Mou JL, Chen Y, Wang MS, Chen Y, Hong Z. An efficient synthesis of oxime compounds catalyzed by superfine kaolin under microwave irradiation. Chinese Journal of Organic Chemistry. 2019;**39**(5):1323-1332

[34] Boukachabia M, Guezane-Lakoud S, Bendjeffal H, Haffas M. Box–Behnken design for the optimisation of Kabachnik–fields reaction catalysed by natural kaolinite clay under eco-friendly conditions. Reaction Chemistry & Engineering. 2024;**9**(5):1164-1172

[35] Pienkoß F, Ochoa-Hernandez C, Theyssen N, Leitner W. Kaolin: A natural low-cost material as catalyst for isomerization of glucose to fructose. ACS Sustainable Chemistry & Engineering. 2018;**6**:8782-8789

[36] Larba SE, Boukachabia M, Zeror S, Bendjeffal H, Bouaroudj T. Synthesis of terpenic esters using natural kaolinite clay: Thermodynamic, kinetic and bioinformatics studies. Flavour and Fragrance Journal. 2025;**40**(4):711-725

[37] Mirjalili BBF, Soltani R. Nano-kaolin/Ti^{4+}/Fe_3O_4: A magnetic reusable nano-catalyst for the synthesis of pyrimido[2,1-b]benzothiazoles. RSC Advances. 2019;**9**(33):18720-18727

[38] Stiti MZ, Habila T, Khaled A, Bouhedja M, Pireaux JJ, Khelili S. A highly efficient and green method for catalyzing the Hantzsch reaction under solvent-free conditions using a seashell/kaolin calcined system. Journal of Chemical Sciences. 2023;**135**(3):70

[39] Wang YD, Liu J, El-Kott AF, AlSheri AS, Ghamry HI. Curcumin-mediated synthesis of silver nanoparticles immobilized on chitosan-modified kaolin: Investigation of its catalytic activity, antioxidant and anti-lung cancer effects. International Journal of Biological Macromolecules. 2025;**307**(1):141540

[40] Gordi Z, Eshghi S. Application of natural kaolin supported sulfuric acid as an ecofriendly catalyst for the efficient synthesis of bis(indolyl) methanes. Synthesis and Reactivity in Inorganic and Metal-Organic Chemistry. 2012;**42**(7):905-908

[41] Sidorenko AY, Khalimonyuk TV, Mamatkodirov BD, Yakubov YY, Aho A, Sviridova TV, et al. Preparation, acid modification and catalytic activity of kaolinite nanotubes in α-pinene oxide isomerization. RSC Advances. 2024;**14**(34):25079-25092

[42] Nasrollahzadeh M, Shafiei N, Baran T, Pakzad K, Tahsili MR, Baran NY, et al. Facile synthesis of Pd nanoparticles supported on a novel Schiff base modified chitosan-kaolin: Antibacterial and catalytic activities in sonogashira coupling reaction. Journal of Organometallic Chemistry. 2021;**945**:121849

[43] Baran T, Nasrollahzadeh M. Facile fabrication of magnetically separable palladium nanoparticles supported on modified kaolin as a highly active heterogeneous catalyst for Suzuki coupling reactions. Journal of Physics and Chemistry of Solids. 2020;**146**(2020):109566

[44] Shahriari M, Sedigh MAH, Shahriari M, Stenzel M, Zangeneh MM,

Zangeneh A, et al. Palladium nanoparticles decorated chitosan-pectin modified kaolin: It's catalytic activity for Suzuki-Miyaura coupling reaction, reduction of the 4-nitrophenol, and treatment of lung cancer. Inorganic Chemistry Communications. 2022;**141**:109523

[45] Bao U, Muschin T, Bao A, Bao YS, Jia M. FeNP-loaded coal-bearing kaolin catalysts for the direct esterification of benzoic acid with cyclicether via C(sp^3)-H bond activation. Green Chemistry Letters and Reviews. 2021;**14**(4):565-577

[46] Gan DF, Wang ZY, Li XD, Zhou JZ, Dai B, Yang L, et al. Green synthesis of bimetallic Pd-Ag alloy nanoparticles supported on polydopamine-functionalized kaolin for catalytic reduction of 4-nitrophenol and organic dyes. Applied Clay Science. 2023;**244**(2023):107091

[47] Asmare ZG, Aragaw BA, Atlabachew M. Facile synthesis of natural kaolin-based CuO catalyst: An efficient heterogeneous catalyst for the catalytic reduction of 4-nitrophenol. ACS Omega. 2024;**9**(49):48014-48031

Section 3

Structural Applications and Geopolymers

Chapter 5

Clay-Based Geopolymers Cement for Oil Wells: A Bibliometric Analysis and Literature Review

Barima Money, Siti Qurratu' Aini Binti Mahat, Norasyikin Ismail, Rayan Hassan Modather, David Abutu and Agi Augustine Aja

Abstract

The use of geopolymers, particularly clay-based geopolymer cement (CBG), in oil well cementing has gained increasing attention due to their promising mechanical properties, environmental benefits, and cost efficiency. This chapter presents a bibliometric analysis and comprehensive literature review on the application of CBG in oil well operations. The bibliometric analysis identifies key research trends, influential publications, leading authors, and collaborations within this field. The literature review focuses on the performance of CBG under high-pressure, high-temperature (HPHT) conditions, their thermal stability, and resistance to chemical degradation in the aggressive environments typical of oil wells. Additionally, the potential of clay-based geopolymers as an eco-friendly alternative to conventional Portland cement is explored, emphasizing their lower carbon footprint, durability, and resource availability. The findings reveal a growing body of work supporting the suitability of geopolymers for oil well cementing applications while highlighting areas requiring further investigation, such as long-term durability and large-scale field applications. This review provides a valuable resource for researchers and industry professionals seeking to develop sustainable cementing solutions for oil and gas wells.

Keywords: geopolymer, oil well cementing, bibliometric analysis, high pressure high temperature (HPHT), environmental sustainability, mechanical properties

1. Introduction

The oil and gas industry faces increasing demands for safer, more efficient, and environmentally sustainable operations, particularly in drilling and well cementing operations [1, 2]. Ordinary Portland cement (OPC) has long been the primary choice for wellbore cementing due to its availability and relatively low cost. However, its performance in harsh downhole environments, particularly under HPHT conditions, poses significant challenges [3]. OPC tends to experience durability issues such as cracking, shrinkage, and chemical degradation, which can compromise the structural integrity of the wellbore. These limitations have driven research into alternative materials capable of overcoming these deficiencies.

In recent years, geopolymers, aluminosilicate materials made from industrial by-products like fly ash, slag, and clays, have emerged as promising alternatives to Portland cement in oil well cementing. Clay-based geopolymers (CBG) offer advantages such as higher thermal stability, chemical resistance, and a lower carbon footprint, making them ideal for HPHT (high pressure, high temperature) environments. They also provide superior mechanical properties, including high compressive strength and durability, while reducing CO_2 emissions [4]. Natural clays like kaolinite, bentonite, red mud, and metakaolin are abundant and low-cost and can be activated into durable materials resistant to chemical attack and thermal cycling in harsh well conditions. Though widely used in construction, geopolymer application in oil and gas remains limited. Studies such as Ref. [5] demonstrated that class F fly ash geopolymer achieves longer thickening times and gains strength over time, unlike Portland cement, which may weaken. Ahdaya and Imqam [6], showed geopolymer's superior rheological and mechanical properties versus OPC. Clays require calcination (e.g., metakaolin) to become reactive; Adjei [4], found elevated curing temperatures (163°F, 248°F) induce new crystalline phases in metakaolin that may cause thermal stress and weaken the system, impacting well integrity.

This chapter presents a bibliometric analysis and comprehensive review of clay-based geopolymers in the context of oil well cementing. The bibliometric analysis aims to uncover the trends and collaborations in this evolving research field, highlighting key authors, publications, and institutions leading the way. Simultaneously, the literature review examines the performance of clay-based geopolymers under HPHT conditions, their thermal stability, and chemical resistance, providing a holistic view of their potential as a sustainable and efficient solution for well cementing applications.

2. Significance of study

The oil and gas industry operates under extreme downhole conditions characterized by high temperatures, elevated pressures, and corrosive chemical environments, factors that often compromise the performance and integrity of conventional Portland cement. In response, this chapter investigates clay-based geopolymer (CBG) cement as a promising alternative, offering superior thermal stability, chemical resistance, and mechanical durability, which can significantly enhance well integrity and reduce the frequency of cement-related failures. Beyond performance, CBG cement presents an economically viable solution, leveraging abundant and low-cost materials, such as industrial and agricultural wastes, thereby reducing both production costs and long-term maintenance expenditures. To gain a comprehensive understanding of the evolving scientific landscape, a bibliometric (scientometric) analysis was employed to systematically map global research trends, key contributors, prolific institutions, and dominant publication sources in the domain of geopolymer cement, with a specific focus on CBG systems. The analysis reveals a steadily increasing interest in CBG cement, driven by the global push for sustainable materials and environmental compliance in energy-intensive sectors. However, the bibliometric mapping also uncovers several critical gaps in the current research ecosystem:

- Limited field-scale validation: Most studies remain laboratory-based, with few scaled-up field trials or long-term performance evaluations in real oil well conditions.

- Fragmented collaboration networks: There is a noticeable lack of interdisciplinary and international collaborations, which hinders knowledge integration and innovation across regions.

- Underexplored formulation optimization: While many studies focus on raw material characterization, fewer delve into the synergistic effects of additives such as magnesium, sodium silicate, or bio-based activators on performance and setting behavior.

- Insufficient standardization: A lack of unified testing protocols and performance benchmarks makes it difficult to compare findings across studies or promote industrial uptake.

- Geographical research imbalance: Research is heavily concentrated in specific countries, with limited contributions from regions rich in untapped natural clay or biomass waste resources.

The study not only identifies these gaps but also outlines strategic future directions, such as developing standardized testing frameworks, fostering multi-sector partnerships, and exploring novel waste-derived activators for sustainable CBG cement production. Furthermore, the valorization of industrial by-products like fly ash, slag, and empty fruit bunch ash (EFBA) reinforces circular economy principles by minimizing landfill disposal and carbon footprints. By integrating performance evaluation with scientometric insights, this work provides a dual contribution, advancing the technical understanding of CBG cement while guiding research policy, funding priorities, and collaborative strategies for broader adoption in oil well cementing and beyond.

3. Methodology

To systematically assess the evolution, research impact, and thematic trends related to CBG cement in oil well applications, a bibliometric analysis was conducted. This quantitative method enables the evaluation of scientific output, identification of influential authors and institutions, and visualization of knowledge structures and research networks within a field.

The Scopus® database was selected as the primary data source due to its broad coverage, reliability, and indexing of high-impact peer-reviewed literature. The search was performed in August 2024, using a strategic combination of keywords: *"geopolymers," "clay,"* and *"oil well cementing."* These terms were selected to capture literature at the intersection of geopolymer science, clay materials, and oil well cementing technologies.

The initial search results were exported in CSV format, including full bibliographic details such as titles, abstracts, author information, affiliations, source journals, keywords, and citations. The dataset was then imported into VOSviewer (version 1.6.20), a widely accepted, open-source software for constructing and visualizing bibliometric networks [5]. This tool was employed to generate coauthorship networks, keyword co-occurrence maps, citation analysis, and journal impact clusters. To ensure the relevance and quality of the dataset, a rigorous data refinement process was applied. This involved the removal of duplicates, non-English records, conference abstracts, and documents unrelated to the target research scope. **Table 1** summarizes the total number of documents retrieved, while **Table 2** outlines the step-by-step filtering criteria used to exclude irrelevant or low-quality publications.

S/N	Keywords searched	Article results	Article results after applying limits
1	Geopolymer	14,459	3367
2	Clay	271,609	32,604
3	Oil well cementing	7085	456

Table 1.
Searched keywords and subsequent documents from the Scopus® database as of August 2024.

Options	Limits and filters applied
Subject area	Earth and Planetary Sciences Environmental Science Engineering Materials Science Chemical Engineering Multidisciplinary
Document type	Articles and reviews
Source type	Journal
Language	English language
Open access	All open access

Table 2.
Limits and filters are applied during data retrieval from the Scopus® database.

By applying this methodology, the study provides a clear, data-driven perspective on the research landscape of CBG cement, revealing knowledge gaps, emerging hotspots, and opportunities for interdisciplinary collaboration.

The retrieved CSV files were imported into VOSviewer for analysis, ensuring data consistency and homogeneity. The study evaluated publication sources, keyword co-occurrence, authors, documents, and countries. **Table 3** lists the settings and limits used in VOSviewer. Maps were created to visualize networks and co-occurrences, while tables presented statistical data. **Figure 1** outlines the scientometric assessment process used in this study.

Analysis type	Selections	Limit applied	Number of thresholds
Mapping of publication sources	Analysis type, analysis unit, minimum publications.	Bibliographic coupling, sources, 3.	43 meet the threshold
Keywords co-occurrence	Analysis type, analysis unit, minimum co-occurrence.	Co-occurrence, all keywords, 12.	98 meet the threshold
Authors mapping	Analysis type, analysis unit, minimum publications.	Co-authorships, authors, 3.	85 meet the threshold
Articles mapping	Analysis type, analysis unit, minimum citations.	Bibliographic coupling, documents, 26.	100 meet the threshold
Countries mapping	Analysis type, analysis unit, minimum publications.	Bibliographic coupling, countries, 2.	42 meet the threshold

Table 3.
Various selections were selected and constraints applied during VOSviewer analysis.

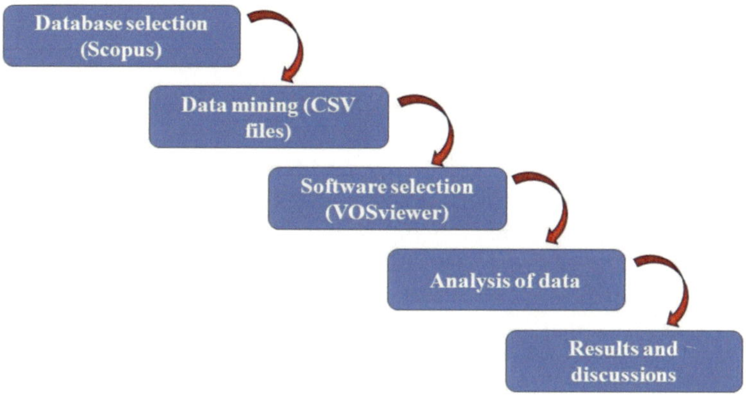

Figure 1.
The scientometric analysis sequence was followed in the present study.

4. Yearly publication trend

Figure 2 shows the yearly publication trends based on three keywords: “geopolymers,” “clay,” and “oil well cementing,” as listed in **Table 1**. For “geopolymers,” the first article appeared in 2000, with only one paper up to 2004. Growth was gradual until 2010, then surged to 3301 publications by August 2024. A similar trend was seen with “clay,” which began in 2000 with 139 publications, rising sharply to 3946 by 2010 and reaching 28,519 by August 2024, indicating strong interest in clay-based, eco-friendly, cost-effective materials. “Oil well cementing” had slower growth, starting in 2001 with just 2 publications, then increasing to 378 by August 2024. Among the three, “clay” showed the most rapid growth, highlighting its importance in geopolymer cement development.

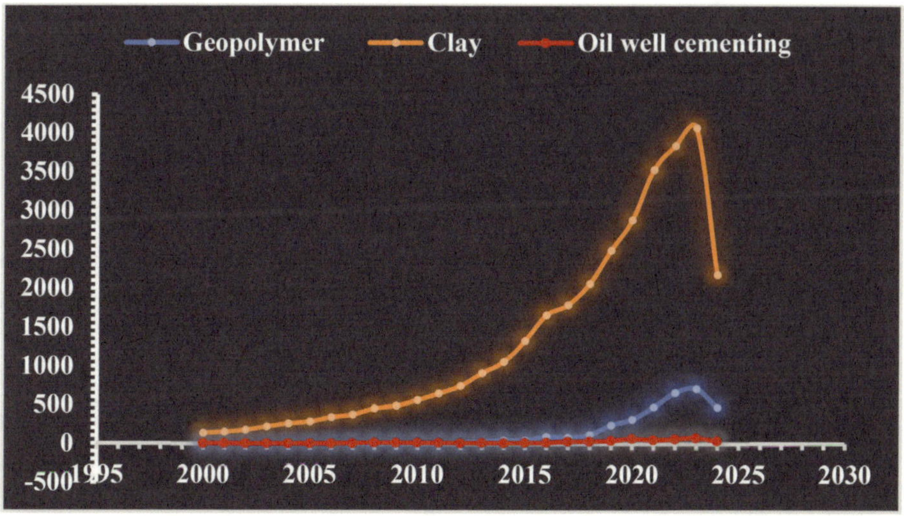

Figure 2.
Yearly publication inclination of the current study area.

4.1 Mapping of publication sources

Mapping publication sources helps track research progress and innovation. **Table 4** lists the top 10 journals publishing on CBG using the keywords "geopolymer," "clay," and "oil well cementing," along with citation counts and h-indexes. *Energy*, *Journal of Petroleum Science and Engineering*, and *Cement and Concrete Research* are the top three, with 42, 34, and 24 papers, respectively. *Cement and Concrete Research* leads in impact with 1791 citations and an h-index of 288 due to rising interest in sustainability and biomass. The *Journal of Petroleum Science and Engineering*, a Q1 journal with 953 citations and an h-index of 148, is popular despite its low acceptance rate due to its high standards. **Figure 3** shows the overlay visualization of journals publishing at least 15 CBG-related papers. Larger circles represent greater influence, with top journals like *Energy* and *Cement and Concrete Research* standing out. VOSviewer identified three clusters: Group 1 (gray/purple) with 7 items, Group 2 (forest green) with 15, and Group 3 (yellow) with 11. Journals in the same cluster are more closely connected. For example, *Energy* is strongly linked to *Geo-Energy Science and Engineering* (forest green), while *Cement and Concrete Research* is linked to *Advanced Concrete Technology* (gray/purple), and *Journal of Petroleum Science* connects with *Frontiers in Energy Research* (yellow).

4.2 Keywords co-occurrence

Co-occurrence networks visualize how often terms appear together in research, helping identify relationships and clusters of related concepts [7]. Keywords play a key role in highlighting core research areas [8]. **Table 5** lists the 16 most frequent keywords, with VOSviewer identifying "cement," "oil well cementing," "oil wells," "compressive strength," and "oil field equipment" as the top five. In **Figure 4(a)**, larger keyword circles indicate higher frequency, and clustering shows keyword groupings. Five clusters were found in red, green, blue, and yellow, with varying keyword counts. **Figure 4(b)** shows keyword density in red, indicating high occurrence, helping future researchers select impactful keywords for better literature access.

S/N	Sources	Document	Citation	H-index
1	Energy	42	332	152
2	Journal of petroleum science and engineering	34	953	148
3	Cement and concrete research	24	1791	288
4	Journal of petroleum exploration and engineering	19	289	46
5	Materials	12	173	640
6	Petroleum science	11	24	148
7	Processes	11	15	182
8	Frontiers in materials	10	17	82
9	Natural gas industry	10	179	28
10	Oil and gas science and technology	9	214	21

Table 4.
List of the top 10 sources of publications, citations, and h-index based on the three keywords from 2000 to August 2024.

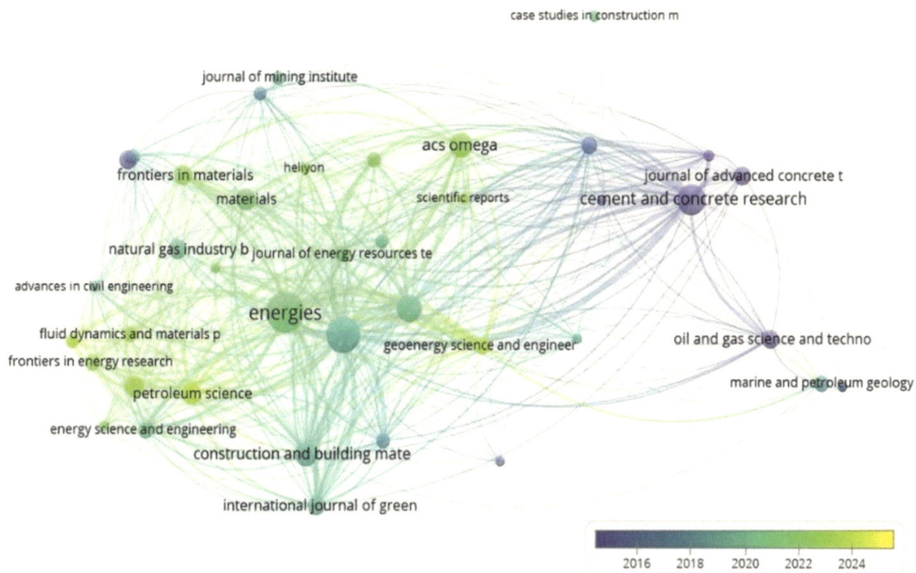

Figure 3.
Overlay visualization of top publication sources.

4.3 Authors mapping

A researcher's influence in a field is often gauged by their citation count. **Table 6** lists the top authors in CBG research by number of publications and citations. Citation averages were calculated by dividing total citations by total publications per author. Although Mahmoud Khalifeh had several publications, his citation count was lower than other leading researchers. Author impact is best assessed by considering multiple metrics. In terms of publication volume, the top contributors were Hajiabadi et al. [5] with 14 papers and 312 citations, Adjei et al. [9] with 13 papers and 1607 citations, and Wu et al. [10] with 11 papers and 59 citations. When ranked by total citations, Adjei et al. [9] led with 1607, followed by Shafiq et al. [11] with 889, and Bu et al. [12] with 439. Average citations per article were as follows: Ref. [11] ~127, Ref. [9] ~124, Ref. [12] ~55, Ref. [13] ~45, Ref. [14] ~33, and Ref. [15] ~28. **Figure 5** presents a citation-based mapping of authors with at least eight publications, generated through bibliometric analysis. The visualization was constructed using settings such as attraction-repulsion layout, 60° horizontal rotation, clustering resolution of 1.00, and association strength normalization.

4.4 Articles mapping

The number of citations an article receives indicates its impact within a research field, with highly cited articles regarded as pioneering works. **Table 7** lists the most cited articles in the field of clay-based geopolymer (CBG) cement, each with at least 100 citations, alongside their authors, publication year, and citation counts. Notably, Jafariesfad et al. [29] received 340 citations for their article titled "Cement sheath modification using nanomaterials for long-term zonal isolation of oil wells," published in the *Journal of Petroleum Science and Engineering*, a Q1 journal known for high-quality publications. Similarly, Lura et al. [30] and Sun et al. [31] garnered 289 and 255

S/N	Keywords	Occurrences
1	Cements	264
2	Oil well cementing	256
3	Oil wells	153
4	Compressive strength	105
5	Oil field equipment	68
6	Cementing (shafts)	67
7	Hydration	60
8	Boreholes	55
9	Cement slurry	47
10	Cement sheath	45
11	Oil well cement	42
12	Portland cement	38
13	Cement paste	38
14	Porosity	35
15	Wellbore	35
16	Mechanical properties	34

Table 5.
The top 16 most frequently occurring keywords in literature.

citations, respectively, for their papers "Cracking in cement paste induced by autogenous shrinking" and "Rheology, curing temperature and mechanical performance of oil well cement; combined effect of cellulose nanofibers and graphene nano-platelets," published in *Materials and Structures/Constructions* and *Materials and Design*, both Q1 journals. These three articles rank as the top-cited works in this domain.

Additionally, Vipulanandan and Mohammed [19], Wang et al. [20], and de Paula et al. [21] earned 250, 135, and 132 citations, respectively, for their papers on "Smart cement rheological and piezoresistive behavior for oil well application," "Three-dimensional analysis of cement sheath integrity around wellbores," and "Mechanical and rheological behavior of oil well cement slurries produced with clinker containing carbon nanotubes." All were published in the *Journal of Petroleum Science and Engineering*, reinforcing the journal's reputation for quality research. **Figure 6** presents a visualization of the most influential linked articles. **Figure 6(a)** shows a network visualization of citation-linked papers, while **Figure 6(b)** displays a density visualization highlighting areas of higher citation concentration. According to VOSviewer analysis, 85 out of 138 articles are interconnected through citations. In the network map (**Figure 6a**), the proximity of documents reflects their citation relationships, while the density map (**Figure 6b**) indicates clusters of highly cited papers, marking the most influential research within the CBG field.

4.5 Countries mapping

A detailed country mapping was conducted to further contribute to the understanding of research activity in the field of clay-based geopolymer (CBG) cement. This scientific mapping of countries was designed to help readers identify the primary

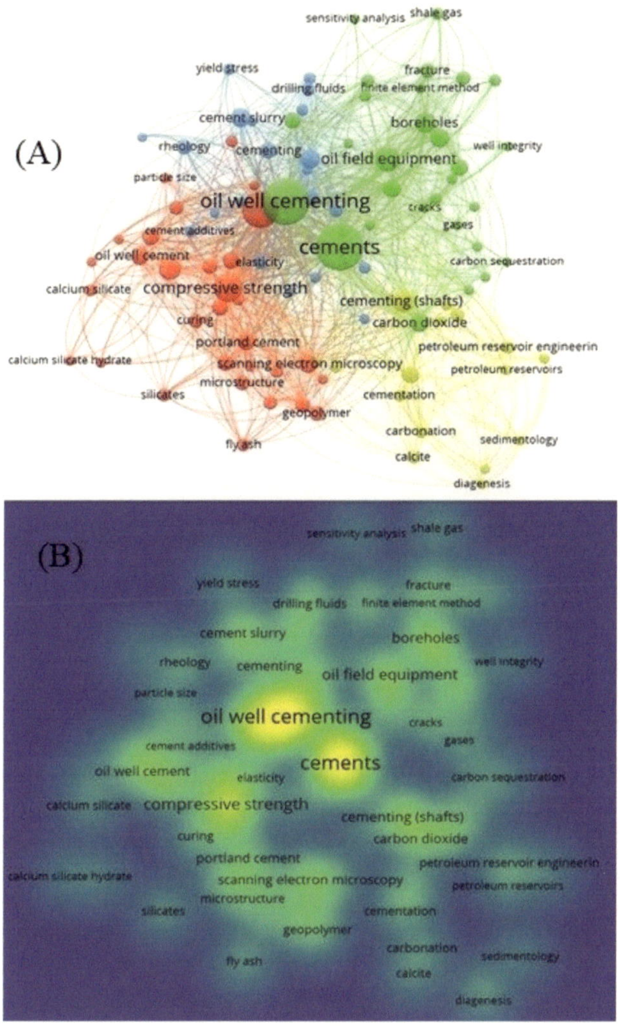

Figure 4.
Co-occurrences of keywords. (A) Network visualization (B) Destiny visualization.

S/N	Author	Documents	Citation	Average citation
1	Khalifeh, Mahmoud	14	312	22
2	Elkatatny, salaheldin	13	1607	124
3	Wu, zhiqiang	11	59	5
4	Cheng, xiaowei	10	437	44
5	Mahmoud, ahmed abdulhamid	9	297	33
6	Li, zaoyuan	9	190	21
7	Xu, mingbiano	9	256	28
8	Ghabezloo, siavash	8	272	34
9	Bu, yuhuan	8	439	55

Table 6.
Authors with a minimum of 8 publications in the current study area.

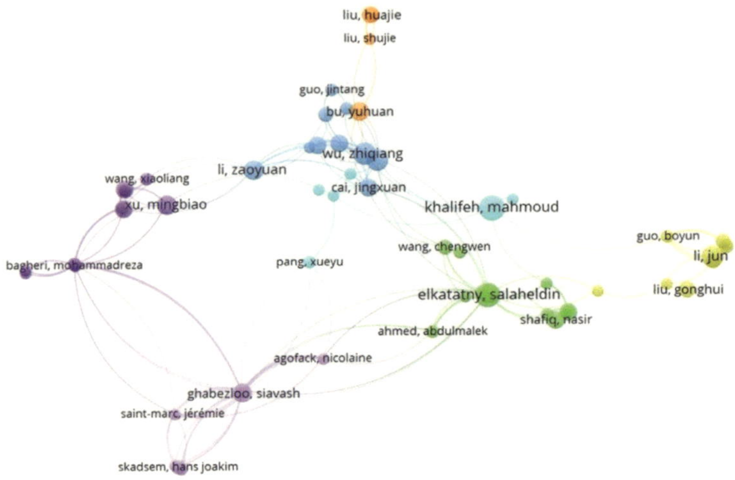

Figure 5.
Science mapping of connected authors based on their citations.

S/N	Articles/ Publication year	Title	Source	Citations
1	[16]	Cement sheath modification using nanomaterials for long-term zonal isolation of oil wells	Journal of petroleum science and engineering	340
2	[17]	Cracking in cement paste is induced by autogenous shrinking	Materials and structure/ constructions	289
3	[18]	Rheology, curing temperature, and mechanical performance of oil well cement; combined effect of cellulose nanofibers and graphene nano-platelets	Materials and design	255
4	[19]	Smart cement rheological and piezoresistive behavior for oil well application	Journal of petroleum science and engineering	250
5	[20]	Three-dimensional analysis of cement sheath integrity around wellbores	Journal of petroleum science and engineering	138
6	[21]	Mechanical and rheological behavior of oil well cement slurries produced with clickers containing carbon nanotubes	Journal of petroleum science and engineering	132
7	[22]	Possibilities of using the ultrasonic wave transmission method to estimate initial setting time of cement paste	Cement and concrete research	125
8	[23]	Influence of nanoclay content on cement matrix for oil wells subjected to cyclic steam injection	Materials	121
9	[24]	Poromechanical behavior of hardened cement paste under isotropic loading	Cement and concrete research	121

S/N	Articles/ Publication year	Title	Source	Citations
10	[25]	Influence of olivine nanosilica on hydration and performance of oil well cement slurries	Materials and design	210
11	[26]	The effect of undrained heating on a fluid-saturated hardened cement paste	Cement and concrete research	115
12	[27]	Oil well cement as an alternative in CO_2-rich environments	Construction and building materials	109
13	[28]	Effect of thermal stress on wellbore integrity during CO_2 injection	International journal for greenhouse gas control	103

Table 7.
Relevant articles with a minimum of 100 citations.

locations by actively researching CBG. The analysis revealed that most published papers originated from authors affiliated with institutions in China, followed by the United States and Norway, making these the top three countries in terms of research output. **Table 8** summarizes the nations with the highest contributions based on the number of published articles and citations related to the current study topic. China led with 202 publications, followed by the United States with 78, and Norway with 34 papers. In terms of citations, the United States ranked highest with 2405 citations, followed by France with 1883 and China with 1811 citations. The United States' leading

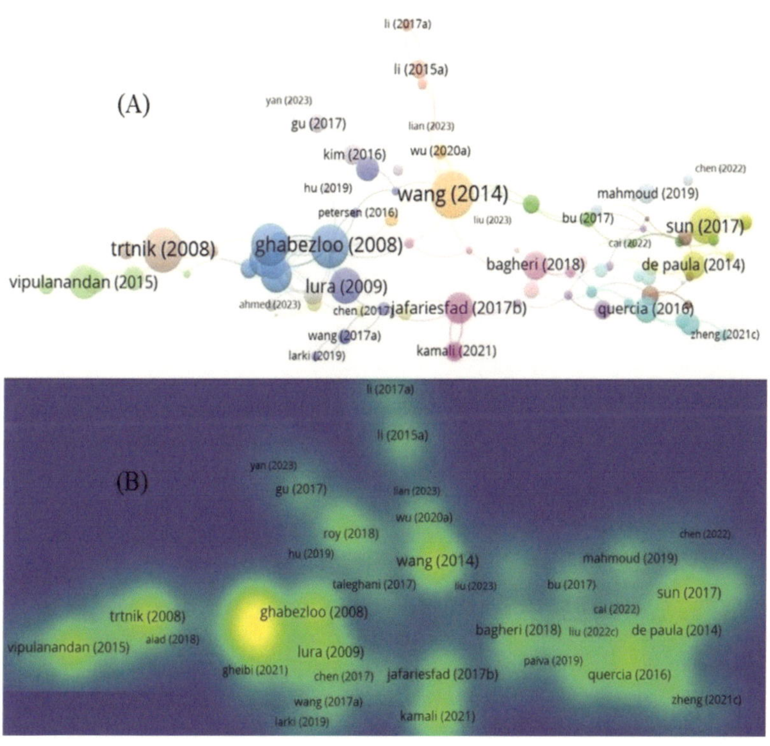

Figure 6.
Articles mapping on publications: (A) Network image of linked papers; (B) Density image of linked papers.

N/S	Country	Documents	Citation
1	China	202	1811
2	United States	78	2405
3	Norway	34	383
4	France	33	1883
5	United Kingdom	29	555
6	Brazil	25	410
7	Malaysia	17	179
8	Saudi Arabia	17	199
9	Japan	16	1119
10	Iran	16	55
11	Russian Federation	15	97
12	Canada	14	280
13	Germany	11	198

Table 8.
The top contributing countries in terms of publications and citations.

citation count is attributed to its strong collaborations with other countries and frequent publications in high-impact journals. The overall publication and citation counts reflect a country's influence and impact in the specific research area. Therefore, it was concluded that China, the United States, Norway, and France have the most significant impact on the current CBG research field. **Figure 7** visualizes the citation-based linkages between countries, where the size of each country's box indicates its level of contribution to CBG research. This graphical representation of active countries is valuable for researchers by fostering scientific collaborations, enabling joint research efforts, and facilitating the exchange of innovative ideas and technologies.

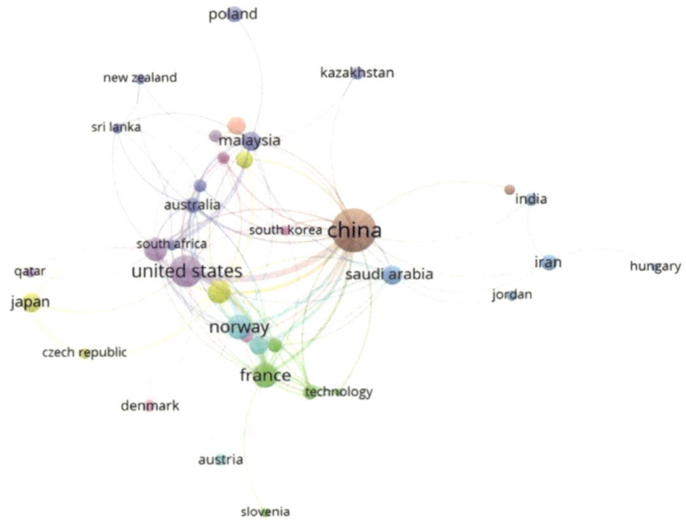

Figure 7.
Network visualization of the leading contributing countries to the current study area.

5. Findings and discussions

A scientometric analysis of CBG cement literature identified key publication sources, common keywords, influential authors, and leading countries in the field [32]. The study found that research on CBG cement is still in its early stages compared to OPC, with more exploration needed, especially for oil well and construction applications. Despite growing interest and progress in the past decade, gaps remain between research and real-world implementation. The review also highlights CBG's potential as an alternative cement and provides insights into its reaction mechanisms in high-temperature oil wells.

5.1 Bibliometric insights

The bibliometric analysis identified significant growth in global interest toward CBG research over the last two decades. Publications on the topic have increased markedly since 2010, reflecting a broader shift toward sustainable construction materials. The keyword trend analysis revealed growing attention to terms like "compressive strength," "oil well cementing," "thermal stability," and "alkaline activation," highlighting the evolving scientific focus on the performance and optimization of CBG systems under harsh conditions. Key journals publishing CBG-related research include *Cement and Concrete Research, Journal of Petroleum Science and Engineering, and Energy*, with strong citation influence, indicating their central role in advancing this field. Author and institution mapping showed contributions primarily from Asia, Europe, and the Middle East but also exposed fragmented collaboration networks, suggesting a need for more international and interdisciplinary partnerships **Table 9**.

5.2 Production of clay-based geopolymer

Geopolymer combines "geo" (referring to aluminosilicate-rich materials) and "polymer" (chains formed from molecular units). CBG materials like kaolin, metakaolin, calcined clay, bentonite, FA, GGBS, silica fume, and various ashes serve as sources of Si and Al. When mixed with alkaline activators, they form a geopolymer binder. CBG is gaining traction for its sustainability, requiring 70% less energy and emitting 90% less CO_2 than OPC [33]. With growing research, CBG shows strong

Natural raw materials	Waste material municipal waste	Industrial waste	Agricultural waste
Kaolin clay	Plastic waste	Fly ash	Rice husk ash
Calcined clay	Rubber waste	Silica fume	Corncob ash
Metakaolin clay	Glass wool fiber	Slag	Palm oil fuel ash
Bentonite clay	Paper waste	Red mud	Coconut husk ash
Volcanic ash	Effluent waste	Industrial recycle glass	Sawdust ash
Zeolite	Cow dung ash	Waste foundry sand	
Dolomite		Tailings	

Table 9.
Types of raw materials used in CBG cement [33].

potential as a green alternative for construction and oil well cementing, especially in high-temperature and zonal isolation applications. **Figures 8** and **9** illustrate its benefits and production process.

CBG cement is a promising alternative to ordinary Portland cement (OPC), utilizing natural clays and industrial by-products to reduce carbon emissions [34].

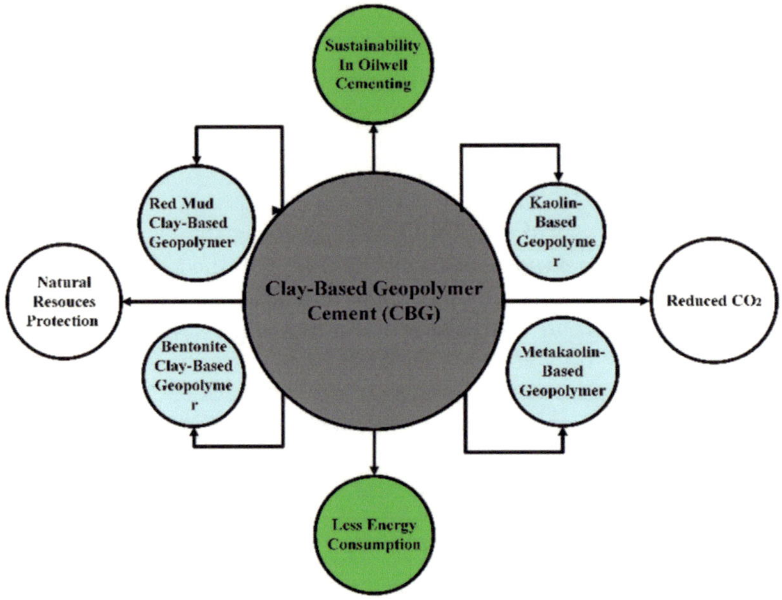

Figure 8.
Schematic representation and the advantages of CBG cement materials.

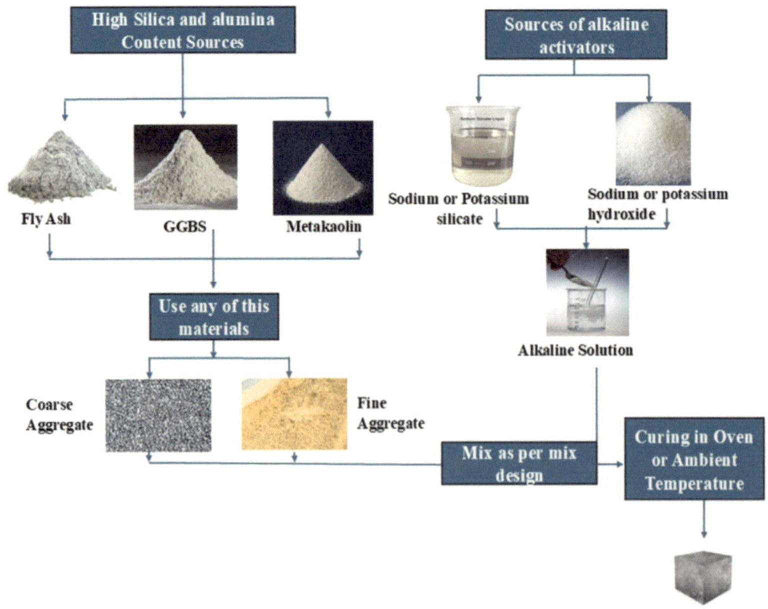

Figure 9.
Process of CBG cement production [4].

DOI: http://dx.doi.org/10.5772/intechopen.1011806

A key component is calcined kaolin, produced by heating kaolin to 700–800°C to enhance reactivity. Even low-reactivity clays can be activated with alkalis like red mud or bentonite to improve performance. According to Yamchelou et al. [35], calcination transforms clay into a reactive form. When mixed with NaOH or KOH, the amorphous Al_2SiO_5 phase dissolves, releasing reactive Si and Al, forming a three-dimensional framework essential for mechanical strength. Adje [4] noted that curing at ambient or elevated temperatures boosts durability and compressive strength. Adding 20% calcined clay to fly ash-based geopolymer improves strength and reduces porosity and shrinkage. Zhang et al. [36] reported excellent resistance of CBG to acid and heat, making it ideal for construction and oil well cementing. Its performance strongly depends on the Si/Al ratio. Studies show metakaolin-based CBG achieves high strength at optimized Si/Al ratios. Duxson et al. [37] tested ratios from 1.15 to 2.15, confirming its critical role in geopolymer performance. **Figure 9** shows the steps involved in the production of CBG cement.

6. Types of clay-based geopolymer

CBGs are a class of inorganic polymers synthesized from aluminosilicate clays activated with alkaline solutions (e.g., NaOH, KOH, and sodium silicate) [38]. The type of clay used, along with processing and curing conditions, influences the final properties of the geopolymer. **Table 10** describes types of CBGs, categorized by clay type and application.

7. Synthesis methods on clay-based geopolymer cement

The synthesis of clay-based geopolymers involves several steps. First, raw clay is dried and ground to reduce moisture and particle size. For kaolinite-rich clays, calcination at 500–800°C produces reactive metakaolin. The clay is then activated with an alkaline solution, usually NaOH (8–16 M), KOH, and Na_2SiO_3, with the silicate-to-hydroxide ratio adjusted to balance workability and strength. Dry ingredients such as clay, fly ash, or empty fruit bunch ash (EFBA) are mixed, followed by the alkaline solution. Mechanical mixing ensures uniformity, and admixtures may be added to modify setting time or flow. The geopolymer paste is then molded or used as a pumpable slurry. Curing typically occurs at ambient temperatures (25–30°C), but elevated curing (60–90°C) or steam/autoclave methods are used to accelerate strength gain, especially for oil well cementing. After curing, samples are demolded and tested for compressive strength, thermal resistance, and chemical stability. Geopolymers may be produced using either a two-part system (mixing solids and liquids before use) or a one-part system (dry mix activated with water on-site). In oil well applications, geopolymerization may occur *in situ* under downhole conditions.

7.1 Reaction mechanisms

Bibliometric analyses of CBG cement research highlight an increasing focus on understanding the fundamental reaction mechanisms that govern geopolymerization, particularly under high-temperature and chemically aggressive conditions relevant to oil well cementing. As synthesized from key literature clusters, the reaction process begins with the alkaline activation of aluminosilicate precursors, primarily kaolin

1. Based on clay type		
Clay type	**Description**	**Common features**
Kaolinite (Kaolin)	Most used, rich in alumina. Often calcined to metakaolin to improve reactivity.	High purity, good geopolymer strength, and excellent pozzolanic reactivity (after calcination).
Montmorillonite (Smectite group)	Layered clay with high cation exchange capacity and swelling behavior.	Requires careful control of water and slower reactivity. Can contribute to a high surface area.
Illite	Potassium-rich mica-type clay with lower reactivity.	May need thermal activation; lower geopolymer strength than kaolin-based types.
Halloysite	Similar to kaolinite but tubular in structure; found naturally or modified.	Promising due to unique morphology and high surface area.
Mixed clay (e.g., natural or bentonite)	It contains combinations of kaolinite, montmorillonite, and others.	Variable reactivity often requires blending or pretreatment.
2. Based on thermal activation		
Activation method	**Clay used**	**Example product**
Calcined clay geopolymers	Kaolin → Metakaolin (500–800°C)	High-performance binders for construction.
Natural clay geopolymers	Raw or partially dried kaolinite, bentonite, etc.	Low-energy geopolymers and eco-friendly construction materials.
3. Based on application		
Application area	**Clay type**	**Key properties desire**
Oil well cementing	Kaolin, bentonite, smectite	High compressive strength, thermal resistance, and low permeability.
Construction materials	Kaolin, illite	Strength, setting time, and durability.
Fire-resistant panels	Metakaolin	High thermal stability and lightweight.
Waste immobilization	Montmorillonite, halloysite	Chemical stability and encapsulation capacity.
3D printing/additive manufacturing	Fine-grained kaolin or halloysite	Rheology control and fast setting.
4. Based on additive material (hybrid geopolymer)		
Type	**Additives**	**Purpose**
Clay + fly ash	Class F fly ash	Improves strength and workability.
Clay + GGBFS	Ground granulated blast furnace slag	Enhances early strength.
Clay + biomass ash (e.g., EFBA)	EFB ash	Utilizes agro-waste and provides silica and alkali.

Table 10.
A summary of types of CBG, activation methods, and application areas.

and metakaolin. These materials undergo dissolution in a high-pH alkaline medium, typically potassium hydroxide (KOH) or sodium hydroxide (NaOH), releasing reactive silicate (SiO_4^{4-}) and aluminate (AlO_4^{5-}) species into solution [39]. Higher temperatures (above 65°C) enhance dissolution by increasing molecular movement, while lower temperatures slow this process. At high pH, the dissolved material forms a supersaturated solution, leading to gel formation as oligomers condense and release water. **Figure 10** shows the reaction mechanism: (A) bibliometric visualization,

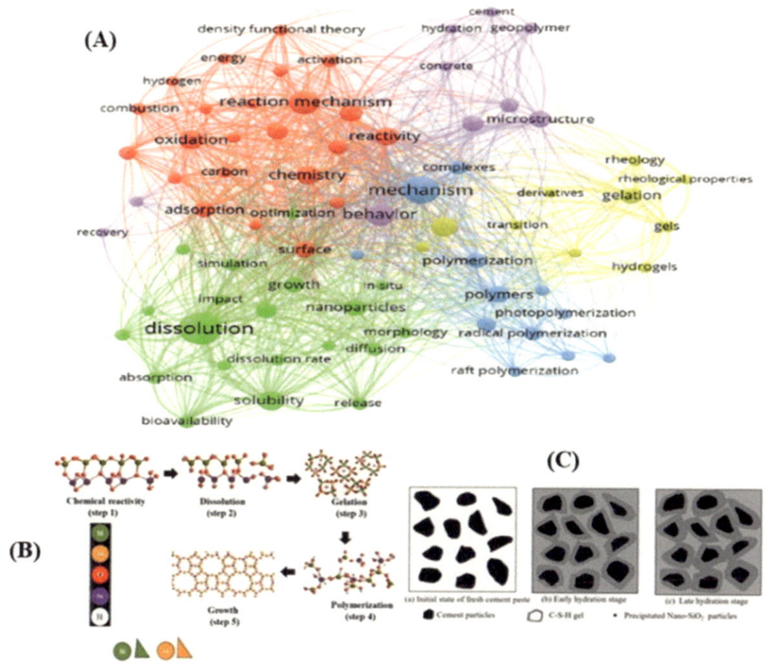

Figure 10.
Reaction mechanism: (A) bibliometric visualization analysis, (B) structure, and (C) microstructure of CBG cement.

showing keyword co-occurrence related to geopolymer structure and curing temperature; (B) schematic of the molecular structure, highlighting Si-O-Al linkages; and (C) microstructural development, displaying the hydration effect on CBG cement.

The bibliometric evidence suggests that temperature, alkalinity, and precursor reactivity are dominant research frontiers in optimizing the CBG reaction pathway. These factors critically influence mechanical strength, setting time, and chemical resistance, aligning with the broader goal of replacing traditional Portland cement in demanding environments.

7.1.1 Dissolution phase

The initial step in the geopolymerization process involves the dissolution of Al_2SiO_5 materials (such as clay) when mixed with an alkaline activator, typically NaOH and Na_2SiO_3. During this phase, the hydroxyl ions (OH^-) from the alkaline solution interact with the Si and Al atoms in the clay structure, breaking covalent bonds Si-O-Si and Al-O-Al, forming reactive silicate (SiO_4^{4-}) and aluminate (AlO_4^{5-}) tetrahedral units [40]. Singh et al. [41] stated that the dissolution is crucial because it creates a pool of reactive species that will participate in subsequent reactions. Eq. (1) shows the chemical reaction that takes place during the dissolution phase. Bibliometric mapping shows a strong correlation between thermal effects and reaction kinetics in highly cited works. Lower temperatures, as noted in less frequently cited studies, result in slower reaction rates and incomplete geopolymer formation.

$$Al_2Si_2O_5(OH)_4 + 4\mathrm{NaOH} + 4\mathrm{H_2O} \rightarrow 2Na_2SiO_3 + 2\mathrm{NaAlO_2} + 6\mathrm{H_2O} \quad (1)$$

7.1.2 Gelation phase

After dissolution, the system enters the gelation phase, where dissolved Si and Al species condense into a gel-like structure through linked tetrahedral units [42]. The Si/Al ratio significantly influences this process; higher ratios typically enhance compressive strength and durability by forming a denser, more interconnected network. As oligomers link up, they create a cohesive gel that gradually hardens as water is lost during curing [43]. This hardening phase results in a dense, three-dimensional network known as the geopolymeric matrix, which is chemically represented in Eqs. (2) and (3).

$$SiO_2 + NaoH \rightarrow Na_2SiO_3 + H_2O \tag{2}$$

$$Al_2O_3 + NaoH \rightarrow NaAlO_2 + H_2O \tag{3}$$

7.1.3 Polycondensation

The final stage is polycondensation, where the gel undergoes further structural rearrangement. The gel transforms into a three-dimensional (3D) network of interconnected Si and Al tetrahedra, forming a hardened geopolymeric structure [44]. This network formation is characterized by strong covalent bonds that provide enhanced mechanical properties compared to traditional Portland cement. The resulting material exhibits high strength, thermal stability, and resistance to chemical attacks [45]. Esposito [46] reported that polycondensation occurs after the initial gelation phase, where hydrolysis and condensation reactions take place. In this phase, the wet gel undergoes structural rearrangement, leading to the formation of a robust polymeric network. This network is characterized by strong covalent bonds that significantly enhance mechanical properties compared to traditional materials like Portland cement. Bibliometric clustering reveals increasing interest in the gelation phase, particularly the transition from sodium or potassium aluminosilicate gel (N-A-S-H or K-A-S-H) to a hardened, durable matrix.

8. Factors influencing geopolymerization

Three basic factors influence the geopolymerization process, which are (1) alkaline activator composition, (2) curing conditions, and (3) raw material characteristics.

- *Alkaline activator composition*: The $NaOH/Na_2SiO_3$ ratio greatly impacts geopolymer reaction kinetics and final properties. NaOH aids in dissolving Al_2SiO_5, releasing Al and Si ions, while Na_2SiO_3 supports gel formation [47]. Higher NaOH speeds up dissolution but can cause rapid gelation, reducing workability. Excess Na_2SiO_3, on the other hand, may slow dissolution and weaken the material. A balanced ratio ensures efficient reactions and optimal strength.

- *Curing conditions*: Temperature and curing time are key to developing desired mechanical properties. Higher temperatures, like 80°C, accelerate reactions and improve strength by promoting stable binding phases [48]. Short curing times at elevated temperatures (6–24 hours) can match or exceed the strength achieved with longer ambient curing. Some clays, such as kaolinite, require thermal activation to react effectively, making heat curing essential for strength development in certain clay-based geopolymers.

- *Raw material characteristics*: The type of clay precursor greatly affects geopolymer performance. Clays such as kaolinite, bentonite, metakaolin, and red mud vary in composition, influencing reactivity, setting time, and strength [49]. Their mineral structure determines how easily they dissolve during activation. Clays with stable crystalline structures need harsher conditions to release reactive Si and Al, impacting the polymerization process.

8.1 Advantages of clay-based geopolymer cement

Because of the unique three-dimensional network structure of the aluminosilicate-based polymer, CBG cement can withstand harsh environments without losing its structural integrity. This property gives CBG cement many great advantages over OPC cement, ceramics, and organic polymers in a variety of application areas [50]. They are described briefly later.

i. *Excellent mechanical properties*: CBG exhibits high mechanical strength due to its robust three-dimensional $[Si\text{-}O\text{-}Al\text{-}O]_n$ network, like ceramics. Unlike traditional Portland cement, which can crack under pressure due to weak van der Waals and hydrogen bonds, CBG's strength is not reliant on such low-strength bonds, resulting in superior load-bearing performance.

ii. *High durability*: CBG is composed of inorganic materials rich in zeolite-like minerals such as sodalite and analcime. These stable frameworks do not suffer from aging like organic polymers and offer superior resistance to extreme environmental conditions compared to Portland cement, contributing to the long-term durability of CBG.

iii. *Low-cost and abundant resources*: CBG is primarily composed of silicon, aluminum, and oxygen elements abundantly available in the Earth's crust. Initially based on kaolin minerals, CBG production now increasingly utilizes industrial waste, making raw materials both cost-effective and widely accessible while promoting resource recovery.

iv. *Special structural applications*: The ring-chain and cage-like structures formed by $Si(OH)_4^-$ and $Al(OH)_4^-$ tetrahedra give CBG unique multifunctional potential. This architecture supports applications in construction, nuclear waste containment, heavy metal immobilization, oil well cementing, and the production of advanced inorganic membranes.

v. *Excellent chemical, fire, and thermal resistance*: CBG's stable Si-O and Al-O bonds offer strong resistance to most acids, excluding hydrofluoric acid. Its oxide network remains intact at high temperatures, and with low thermal conductivity (0.24–0.38 W/(m·K)), CBG also serves effectively as a heat-insulating material.

vi. *Low resource use and CO_2 emissions*: Producing metakaolin requires lower temperatures (500–800°C) than Portland cement (800–1300°C), leading to ~80% energy savings and 50–80% less CO_2 emissions, supporting sustainable development goals.

vii. *Fast curing and strong interfacial bonding*: CBG develops high early strength due to rapid gelation and dehydration. Unlike traditional cement, it forms a stable bond with aggregates without a weak transition zone, making it ideal for concrete repair applications.

9. Application of clay-based geopolymer cement

CBG technology offers strong potential for reusing waste materials, helping to mitigate environmental issues associated with landfilling. CBG-based products have found applications in asbestos-free materials, fire-resistant products, oil well cementing, construction, hazardous waste stabilization, advanced ceramics, and high-performance materials [51]. Cement plays a vital role in modern infrastructure, and geopolymers present a promising alternative by enabling eco-friendly solutions, scalable production, and broader commercialization. For successful implementation, the optimal mix design and processing conditions must align with local raw material availability and geographic factors. **Figure 11** illustrates the bibliometric mapping of CBG cement applications, including its use in zonal isolation, well integrity, extreme condition performance, carbon nanotube incorporation, wellbore stability, soil stabilization, and acid resistance.

In Malaysia, clay-based geopolymer (CBG) cement is increasingly used in infrastructure projects like pavements, water tanks, bridge decks, and precast components [52], indicating its growing acceptance in Asia. Its structural viability is supported by studies, including a 2016 initiative by Japan to assess global CBG use and a RILEM study involving 26 labs to standardize durability testing. However, adoption remains

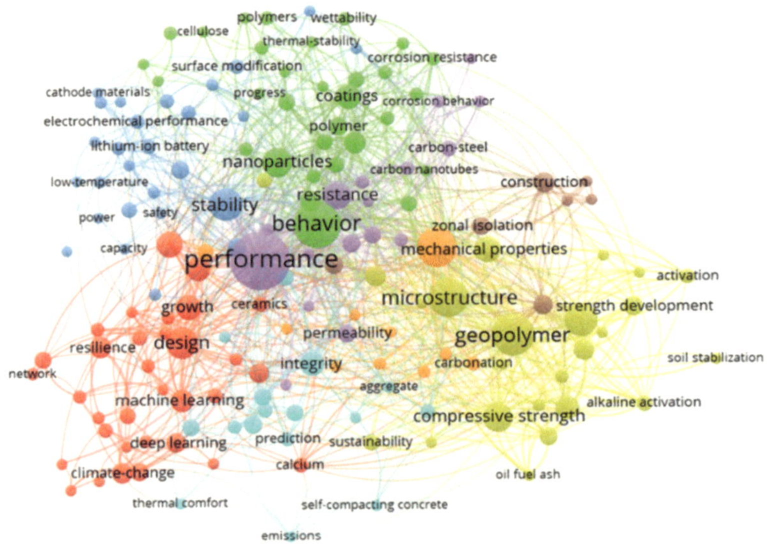

Figure 11.
Bibliometric visualization analysis on the application of CBG cement.

limited due to formulation complexity, curing demands, and supply chain issues. CBG is unlikely to fully replace conventional cement due to these challenges and regulatory hurdles. Still, when produced from local, low-carbon materials, it offers a sustainable alternative for construction and oil well cementing. Its commercial potential is increasingly recognized, and it is also being explored for high-temperature applications such as geopolymer tiles for saturated environments.

9.1 Challenges and prospects

The emergence of CBG cement has opened new possibilities for sustainability in construction and the oil well cementing process because it eliminates the use of ordinary Portland cement as a binder in construction materials. However, certain limitations/restrictions hinder their further applications.

- CBGs often need heat or steam curing for strength, but this can be substituted by using films and a humid environment.

- The use of highly alkaline activators in CBGs can trigger alkali-silica reaction (ASR), especially in high-calcium or mixed systems. ASR risk also depends on activator type and concentration, and some additives may contribute to it. Further research is needed to identify key factors influencing ASR.

- A key challenge in CBG production is the limited availability and high cost of sodium silicate (Na_2SiO_3), which affects both environmental and economic viability. Alternative sources like red mud, bentonite, and waste glass can replace traditional materials such as quartz sand and sodium carbonate, reducing emissions. These waste-based materials also serve as effective and affordable activators for metakaolin-based geopolymers.

- Efflorescence is a challenge in the development of CBG, often caused by excess alkali activators, especially in silicon-rich systems. While higher alumina content helps reduce it, increased silica content and finer silica particles also minimize its occurrence. Techniques like silane surface modification make CBG hydrophobic, limiting water absorption and alkali leaching without affecting strength. Additionally, nanosilica can reduce efflorescence by binding excess alkalis into an amorphous gel.

10. Conclusion

This study presented a comprehensive bibliometric and literature-based evaluation of clay-based geopolymer (CBG) cement, with a focus on its growing potential in oil well cementing and infrastructure applications, particularly within Malaysia and the broader Asia-Pacific region. The bibliometric analysis, derived from the Scopus® database (2000-August 2024) and visualized using VOSviewer, revealed a steady rise in global research interest, driven by CBG's environmental advantages, material performance, and compatibility with industrial waste valorization strategies. From a performance perspective, CBG exhibits outstanding mechanical strength, thermal stability, and chemical resistance, attributes that are especially critical under high-pressure, high-temperature (HPHT) conditions in oil wells. The geopolymerization process, comprising dissolution, gelation, and polycondensation, was found to be highly sensitive to precursor composition, curing temperature, and activator chemistry. Key advantages of CBG include its low-carbon footprint, rapid strength development, and compatibility with locally available or waste-derived materials like EFBA, red mud, and fly ash.

Despite these benefits, several challenges hinder the full-scale adoption of CBG cement. These include the high cost and limited supply of sodium silicate, the need for elevated-temperature curing, risks of alkali-silica reaction (ASR), and efflorescence. In addition, the lack of standardized testing protocols, limited field-scale validation, and fragmented research collaborations represent key gaps identified through bibliometric mapping.

Nonetheless, the prospects for CBG cement are promising. Emerging strategies such as the use of waste-derived activators, low-energy curing techniques, surface modification technologies, and nanomaterial integration are addressing current limitations. The continued development of standardized testing frameworks, coupled with greater interdisciplinary collaboration, will be critical for transitioning CBG from laboratory research to widespread industrial use. CBG cement stands out as a sustainable and high-performance alternative to ordinary Portland cement, particularly for demanding applications such as oil well cementing and environmentally conscious infrastructure development. With the right policy support, technological refinement, and industrial partnerships, CBG has the potential to play a pivotal role in advancing low-carbon construction and circular economy goals globally.

Acknowledgements

This reviewed work was supported by the Ministry of Higher Education (MOHE) Malaysia and Universiti Malaysia Pahang Al-Sultan Abdullah through research grants RDU 220375 and RDU 233305.

Conflict of interest

The authors declare no conflict of interest.

Author details

Barima Money[1*], Siti Qurratu' Aini Binti Mahat[1], Norasyikin Ismail[1], Rayan Hassan Modather[1], David Abutu[1,2] and Agi Augustine Aja[1,3*]

1 Faculty of Chemical and Process Engineering Technology, Universiti Malaysia Pahang Al-Sultan Abdullah, Kuantan, Pahang, Malaysia

2 Department of Chemical Engineering, Federal University Wukari, Nigeria

3 Centre for Research in Advanced Fluid and Processes (Fluid Centre), Universiti Malaysia Pahang Al-Sultan Abdullah, Gambang, Pahang, Malaysia

*Address all correspondence to: moneybarima@gmail.com and augustine@umpsa.edu.my

References

[1] Kanesan D, Irawan S, Ridha S, Chandran D, Azhar NAB. The suitability of fly ash based geopolymer cement for oil well cementing applications: A review. ARPN Journal of Engineering and Applied Sciences. 2018;**13**:8297-8316

[2] Al Ani M, Panis NI, Manikumar Y, Sidek MA, Jaafar MZ, Ridzuan N, et al. Optimization of polymeric nanofluid oil recovery mechanisms at reservoir condition. In: Society of Petroleum Engineers—SPE Nigeria Annual International Conference and Exhibition, NAIC 2024. OnePetro; 2024. DOI: 10.2118/221597-MS

[3] Chen K. Geopolymer concrete durability subjected to aggressive environments—A review of influence factors and comparison with ordinary Portland cement. Construction and Building Materials. 2021;**279**:122496

[4] Adjei S. Geopolymer as the future oil-well cement: A review. Journal of Petroleum Science and Engineering. 2022;**208**:109485

[5] Hajiabadi SH, Khalifeh M, van Noort R. Stability analysis of a granite-based geopolymer sealant for CO_2 geosequestration: In-situ permeability and mechanical behavior while exposed to brine. Cement and Concrete Composites. 2024;**149**:149. SSRN 4684181

[6] Ahdaya M, Imqam A. Investigating geopolymer cement performance in presence of water based drilling fluid. Journal of Petroleum Science and Engineering. 2019;**176**:934

[7] Quercia G, Brouwers HJH, Garnier A, Luke K. Influence of olivine nano-silica on hydration and performance of oil-well cement slurries. Materials and Design. 2016;**96**:162-170. DOI: 10.1016/j.matdes.2016.02.001

[8] Ghabezloo S, Sulem J, Guédon S, Martineau F, Saint-Marc J. Poromechanical behaviour of hardened cement paste under isotropic loading. Cement and Concrete Research. 2008;**38**(12):1424-1437. DOI: 10.1016/j.cemconres.2008.06.007

[9] Adjei S, Elkatatny S, Sarmah P, Chinea G. Investigation of dehydroxylated sodium bentonite as a pozzolanic extender in oil-well cement. SPE Drilling & Completion. 2021;**36**:730

[10] Wu Z, Chen Z, Wang C. Study of retention of drilling fluid layer on annulus wall during cementing. Processes. 2024;**12**(6):1176

[11] Shafiq N, Kumar R, Zahid M, Tufail RF. Effects of modified metakaolin using nano-silica on the mechanical properties and durability of concrete. Materials. 2019;**12**(14):2291

[12] Bu Y, Lu Z, Lu C, Liu H, Guo S, Liu H. Investigating the applicability of novel hydrate dissociation inhibitors in oilwell cement through molecular simulations. Scientific Reports. 2024;**14**(1):14970

[13] Guo X, Pan X. Mechanical properties and mechanisms of fiber reinforced fly ash–steel slag based geopolymer mortar. Construction and Building Materials. 2018;**179**:633-641. DOI: 10.1016/j.conbuildmat.2018.05.198

[14] Ahmed AM, Alzayer BM, Panagopoulos G, Kiomourtzi P, Kirmizakis P, Elkatatny S, et al. A new empirical correlation for pore pressure prediction based on artificial neural

networks applied to a real case study. Processes. 2024;**12**(4):664

[15] Xu L, Wu S, Xu MB, Xu J, Huang ML, Chen ZH. Insights into unique in-situ self-adaptive sealing property of pressure-activated sealant in leaks: Experimental and numerical investigations on mechanism of liquid-to-solid transformation triggered by differential pressure. Geoenergy Science and Engineering. 2024;**234**:212655

[16] Jafari Nadoushan M, Ramezanianpour AA. The effect of type and concentration of activators on flowability and compressive strength of natural pozzolan and slag-based geopolymers. Construction and Building Materials. 2016;**111**:337-347. DOI: 10.1016/j.conbuildmat.2016.02.086

[17] Liew Y-M, Heah C-Y, Mohd Mustafa AB, Kamarudin H. Structure and properties of clay-based geopolymer cements: A review. Progress in Materials Science. 2016;**83**:595-629. DOI: 10.1016/j.pmatsci.2016.08.002

[18] Abdullah MMA, Hussin K, Bnhussain M, Ismail KN, Ibrahim WMW. Mechanism and Chemical Reaction of Fly Ash Geopolymer Cement-a Review. 2011. Available from: www.ijopaasat.in

[19] Vipulanandan C, Mohammed A. Effect of nanoclay on the electrical resistivity and rheological properties of smart and sensing bentonite drilling muds. Journal of Petroleum Science and Engineering. 2015;**130**:86-95. DOI: 10.1016/j.petrol.2015.03.020

[20] Wang C, Chen X, Wang L, Ma H, Wang R. A novel self-generating nitrogen foamed cement: The preparation, evaluation and field application. Journal of Natural Gas Science and Engineering. 2017;**44**:131

[21] de Paula JN et al. Mechanical and rheological behavior of oil-well cement slurries produced with clinker containing carbon nanotubes. Journal of Petroleum Science and Engineering. 2014;**122**:274-279. DOI: 10.1016/j.petrol.2014.07.020

[22] Trtnik G, Turk G, Kavčič F, Bosiljkov VB. Possibilities of using the ultrasonic wave transmission method to estimate initial setting time of cement paste. Cement and Concrete Research. 2008;**38**(11):1336-1342. DOI: 10.1016/j.cemconres.2008.08.003

[23] Mahoud M, Shahhosseini V. Workforce competency models for multiple construction projects: A review. Organization Theory Review. 2019;**3**(2):43-69. journals.umt.edu.pk

[24] Ghabezloo S, Sulem J, Saint-Marc J. The effect of undrained heating on a fluid-saturated hardened cement paste. Cement and Concrete Research. 2009;**39**(1):54-64. DOI: 10.1016/j.cemconres.2008.09.004

[25] Guercia E, Berti F, Navarini L, Demitri N, Forzato C. Isolation and characterization of major diterpenes from C. Canephora roasted coffee oil. Tetrahedron: Asymmetry. 2016;**27**(14-15):649-656. DOI: 10.1016/j.tetasy.2016.06.008

[26] Ghabezloo S. Association of macroscopic laboratory testing and micromechanics modelling for the evaluation of the poroelastic parameters of a hardened cement paste. Cement and Concrete Research. 2010;**40**(8):1197-1210. DOI: 10.1016/j.cemconres.2010.03.016

[27] Bagheri A et al. Microstructural study of environmentally friendly boroaluminosilicate geopolymers. Journal of Cleaner Production.

2018;**189**:805-812. DOI: 10.1016/j.jclepro.2018.04.034

[28] Roy A. Review on the biosurfactants: Properties, types and its applications. Journal of Fundamentals of Renewable Energy and Applications. 2017;**8**(2):10-4172

[29] Jafariesfad N, Geiker MR, Gong Y, Skalle P, Zhang Z, He J. Cement sheath modification using nanomaterials for long-term zonal isolation of oil wells. Journal of Petroleum Science and Engineering. 2017;**156**:662-672

[30] Lura P, Jensen OM, Weiss J. Cracking in cement paste induced by autogenous shrinkage. Materials and Structures/Materiaux et Constructions. 2009;**42**(8):1089-1099. DOI: 10.1617/s11527-008-9445-z

[31] Sun X, Wu Q, Zhang J, Qing Y, Wu Y, Lee S. Rheology, curing temperature and mechanical performance of oil well cement: Combined effect of cellulose nanofibers and graphene nano-platelets. Materials and Design. 2017;**114**:92-101. DOI: 10.1016/j.matdes.2016.10.050

[32] Gheibi S, Agofack N, Sangesland S. Modified discrete element method (MDEM) as a numerical tool for cement sheath integrity in wells. Journal of Petroleum Science and Engineering. 2021;**196**:107720

[33] Bagheri M, Shariatipour SM, Ganjian E. A review of oil well cement alteration in CO_2-Rich environments. Construction and Building Materials. 2018;**186**:946-968

[34] Cai J, Zhou J, Liu C, Mei K, Zhang C, Cheng X. Microencapsulated phase change material-cement composites for cementing the natural gas hydrate layer. Construction and Building Materials. 2023;**399**:132591

[35] Yamchelou MT, Law D, Brkljača R, Gunasekara C, Li J, Patnaikuni I. Geopolymer synthesis using low-grade clays. Construction and Building Materials. 2021;**268**:121066

[36] Zhang B, Fang L, Ding Z, Li M. The effect of modified basalt fiber on mechanical properties of oil well cement slurry. Case Studies in Construction Materials. 2024;**20**:e03228

[37] Duxson P, Fernández-Jiménez A, Provis JL, Lukey GC, Palomo A, Van Deventer JSJ. Geopolymer technology: The current state of the art. Journal of Materials Science. 2007;**42**(9):2917-2933. DOI: 10.1007/s10853-006-0637-z

[38] Goracci G, Ogundiran MB, Barzegar M, Iturrospe A, Arbe A, Dolado JS. Kaolin clay-based geopolymer for ionic thermoelectric energy harvesting. ACS Omega. 2024;**9**(12):13728-13737

[39] Zhu X, Qian H, Wu H, Zhou Q, Feng H, Zeng Q, et al. Early-stage geopolymerization process of metakaolin-based geopolymer. Materials. 2022;**15**(17):6125

[40] Jeyasangar P, Hazwan Hussin M, Ganesan S. Analysis of efflorescence of spent bleaching clay based geopolymer using alkaline leachability methods. Malaysian Journal of Analytical Sciences. 2024;**28**(2):322-334

[41] Singh S, Aswath MU, Ranganath RV. Effect of mechanical activation of red mud on the strength of geopolymer binder. Construction and Building Materials. 2018;**177**:91-101. DOI: 10.1016/j.conbuildmat.2018.05.096

[42] Dahal S. Synthesis of alkali activated geopolymer cement from clay. Journal of Civil and Environmental

Engineering. 2022;**12**:1-13. DOI: 10.37421/2165-784X.22.12.455

[43] Long Q, Zhao Y, Zhang B, Yang H, Luo Z, Li Z, et al. Interfacial behavior of slag, fly ash, and red mud-based geopolymer mortar with concrete substrate: Mechanical properties and microstructure. Buildings. 2024;**14**(3):652

[44] Mustafa AM et al. Chemical reactions in the geopolymerisation process using fly ash-based geopolymer: A review 1. Australian Journal of Basic and Applied Sciences. 2011;**5**(7):1199-1203

[45] Siyal AA, Mohamed RMSR, Shamsuddin R, Ridzuan MB. A comprehensive review of synthesis kinetics and formation mechanism of geopolymers. RSC Advances. 2024;**14**(1):446-462

[46] Esposito S. "Traditional" sol-gel chemistry as a powerful tool for the preparation of supported metal and metal oxide catalysts. Materials. 2019;**12**(4):668

[47] Chen X et al. Utilization of red mud in geopolymer-based pervious concrete with function of adsorption of heavy metal ions. Journal of Cleaner Production. 2019;**207**:789-800. DOI: 10.1016/j.jclepro.2018.09.263

[48] He J, Jie Y, Zhang J, Yu Y, Zhang G. Synthesis and characterization of red mud and rice husk ash-based geopolymer composites. Cement and Concrete Composites. 2013;**37**(1):108-118. DOI: 10.1016/j.cemconcomp.2012.11.010

[49] Bai B, Bai F, Nie Q, Jia X. A high-strength red mud–fly ash geopolymer and the implications of curing temperature. Powder Technology. 2023;**416**:118242

[50] Matsimbe J, Dinka M, Olukanni D, Musonda I. Bibliometric trends of geopolymer research in sub-Saharan Africa. Materials Today Communications. 2023;**35**:106082

[51] Amran M, Al-Fakih A, Chu SH, Fediuk R, Haruna S, Azevedo A, et al. Long-term durability properties of geopolymer concrete: An in-depth review. Case Studies in Construction Materials. 2021;**15**:e00661

[52] Wasim M, Ngo TD, Law D. A state-of-the-art review on the durability of geopolymer concrete for sustainable structures and infrastructure. Construction and Building Materials. 2021;**291**:123381